Go Slow and Curvy

dehyun sohn • Hee-Jung Jang
Timothy Jung

Go Slow and Curvy

Understanding the Philosophy of the Cittaslow^{slowcity} Phenomenon

Springer

dehyun sohn
Cittaslow International
Seoul
Republic of Korea

Hee-Jung Jang
Silla University
Busan
Republic of Korea

Timothy Jung
Manchester Metropolitan University
Manchester
United Kingdom

ISBN 978-3-319-37211-2 ISBN 978-3-319-14854-0 (eBook)
DOI 10.1007/978-3-319-14854-0

Springer Cham Heidelberg New York Dordrecht London

Softcover re-print of the Hardcover 1st edition 2015

Printed on acid-free paper

Springer International Publishing AG Switzerland is part of Springer Science+Business Media (www.springer.com)

Prologue

The base of capital is money, whereas the foundation of "Manbon" is represented by the four cardinal virtues—slowness, freedom, soul, and delight.

Not too long ago, Prof. dehyun sohn, one of the authors received a letter, in which the first paragraph was stating:

> Any pig or cow is able to get up and run once its hair has dried after birth, while humans need as long as 1 year for the legs to gather enough strength in order to attempt the first steps. Following, he will need another 20 years, while receiving family and academic education to become a full human being.

There was a Korean farmer, who Prof. sohn respected due to his gradual lifestyle, for he believed that it is in the roots of human nature to live an unhurried life. Among animals, human beings pass a relatively long infant period, wearing diapers for almost 2 years, while it also takes a long time for babies to walk, falling over 2,000 times in average. However, "slowness" is not simply considered "slow"; in any field it is necessary to look at the past and the present properly in order to better prepare for the future. Even in the Western culture, there is a Latin aphorism "more haste, less speed—festina lente." The same goes for us; the more we live an unhurried life, the more we can consider the past and learn for the future. Without tilting to one side or the other, patience as well as wisdom is necessary in order to find the balance between fast and slow.

South Korea is a country that develops and runs rapidly regardless of being a big, mediocre, or small area. In such a spinning current, we desire a relaxing life. Among such movements, one is internationally called the "Cittaslow" movement. "Slow" is regarded as "joyful"; "Slow" is regarded as "deep." The philosophy of "Cittaslow" is based on matureness instead of development and growth, quality of life instead of quantity of life, and seeks to find the deepness in life instead of speeding through it. Since life needs to have a deep meaning in order to live brighter and happier.

The same goes for a business or for a country. A business or a country can only be successful, if the CEO and politicians have a bright mind, a calm soul, and a peaceful heart. It is not possible to sustain a good economy without the right spirit, nor is this possible with a CEO and politicians who do not seek fulfilness in their heart. Imagine the outcome of a country, whose responsibility is given to a person who is nervous about his actions and constantly under pressure. This has formed the foundation of phenomenon of "slow management," in which slowness, humanity, and nature coexist in a balance to create a brighter future. This slow and steady curve must bear a secret only God himself knows about. Hence, by writing this book, the authors seek to bring the idea of a sustainable "global village" one step closer and propose three "slowwares" including the aspects of "slowness" in Cittaslow, "smallness" of small and medium-sized businesses, and "sustainability" in agriculture. Furthermore, the authors suggest Manbonism, slow capitalism as a sustainable alternative to capitalism.

Generally speaking, the economy must develop slowly but surely. Originally, we were not meant to sacrifice our lives for the economy and capital, but instead make use of the development to increase our living standards. However, it has become inevitable to focus on the economy in order to ensure happiness. In Manbonism, there are four virtues of humanity (slowness, freedom, soul, and delight), while "slowness with curve," "free will," an "ethical and moral soul," and "delight in creation" form the ideal roots of Manbonism. Only if such virtues become the base of our lives, it is possible to create a lasting future and be filled with happiness.

This book is divided into five chapters. The first chapter explains the importance and essence of slowness, smallness, and sustainability while the second chapter deals with the importance of "slowness," covering a slow life for individuals in an emotional society. The third chapter covers the issue of "smallness," especially "slow management" for small businesses and local administration and provides arguments that refer to small businesses as a foundation to true economy and economic democracy, whereas Chap. 4 talks about the love for agriculture.

In Chap. 5, the authors focus on "sustainability," discussing the real meaning of happiness by comparing Capitalism to Manbonism. This is supported by providing various examples as well as with photographs in order to enhance the reader's understanding. The authors themselves are in the process of reaching the true beauty of "slowness" by gradually moving step by step, while bringing peace into his heart and mind, giving every moment in his life a deeper meaning. Although it is not possible to ignore or oppose the power of the "WWW (World Wide Web)" in today's society, it is necessary to combine such new technology with our traditional Korean virtues (pungryu). Thus, by harmonizing and appreciating elements, such as nature, art, and life, everyone should work together building a "WWW (Win-Win-Win)" situation in order to create a community based on humanity and enhance the beauty of this world.

The authors would like to extend their gratitude to Springer for deciding to publish this book, "Go Slow & Curvy," although they are unfamiliar with the theme of the book. It is also honorable to work with Springer, one of the world's leading publishers that has worked with over 150 Nobel Prize winners. It is particularly meaningful for the authors to work with a publisher based in Germany, which is regarded as the saviour of Korea's economic development. Korea was one of the poorest countries with per capital income of 60 USD after Korean War (1950–1953). At that time, West Germany employed many Korean mine workers and nurses and was only nation that granted Korea two commercial loans that became seed money to Korea's economic growth. Finally, the authors really appreciate help of Dai-In Han, Ph.D. researcher of Manchester Metropolitan University and Jude O. Kim, Ph.D. and head researcher of the National Cittaslow Corea Network.

Seoul, Republic of Korea — dehyun sohn
Busan, Republic of Korea — Hee-Jung Jang
Manchester, UK — Timothy Jung
October 2015, 4348

Contents

A Korean cock is an auspicious and domestic poultry announcing the dawn time and preventing disasters, and brining about good fortune. The cock painting was attached on the middle door.

Slow, Small, and Sustainable 1

Contents

1.1 The Secret of Life Hidden in the Slow and Curvy

"Slow" is "deep." Recently, the "slow" movement such as walking, cycling, and other activities based on Mediterranean living has spread significantly [1]. Such vast deceleration in various areas is incorporated in the Cittaslow^Slowcity movement that was started in Italy and means "Slowcity" in English. The movement is a philosophy of the coexistence of "slowness" and "city" that is the epitome of "fast" and promotes living in a free and easy manner in a hectic city as a new urbanism.

The Cittaslow movement states that one should live like a decent human being. Escaping from the competition for "more, faster, and wealthier" in today's society, the movement, based on "slowness, smallness, and sustainability," advises to live "a bit slower, a bit less, and a bit smoother" in order to ensure sustainability.

"Slow" is the philosophy of tilting the weight from growth to maturity and from quantity to quality of lifestyle and thus providing

d. sohn et al., *Go Slow and Curvy*,
DOI 10.1007/978-3-319-14854-0_1

insight into the depth of thought and the mind, human relationships, and work. Although "slow" was regarded rather negative in the twentieth century, "slow" and "small" are highly valued in the twenty-first century. Therefore, it is necessary to share the positive effect of "slowness" among citizens and visitors to destinations, those who want to live in a Cittaslow, and those who want to experience a Cittaslow lifestyle.

The Cittaslow movement does not aim to turn the majority of people from a "fast-based lifestyle" to a "slow-based lifestyle." "Fast" can be thrilling, productive, and powerful. If it were not for the "fast" lifestyle, some countries like Hong Kong and Korea would have remained poor countries until today. However, rather than "faster" and "more," this movement wants to advise that "slow and small are better." The problem hereby is not to reverse time but to find a harmonious rhythm of life between fast and slow, quantity of life and quality of life, rural community and city, local and global, and analog and digital. The movement is the recipe to find the harmony between "la dolce vita (the sweet life)" and the dynamic information society. It suggests moving sometimes fast, sometimes slow, and occasionally with average speed. Our problem is that we are currently unsteadily tending toward being too fast.

Let's think about harmony a bit more. There is a joke about a dental treatment. Once, a young man went to a dentist and asked how much it cost him to have a tooth extracted. As the dentist answered it would cost around USD 300, the young man was shocked and replied, "300 dollars for just a few minutes work, that expensive." The dentist answered with a serious look, "OK. I'll pull it out slowly if you prefer."

The prominent Korean Buddhist scholar, Wonhyo (see Fig. 1.1), said, "When sewing clothes, a short needle is needed and a long spear is useless. To avoid the rain, a small umbrella is needed, and a cover that spans the entire sky is useless. Therefore, small things are appropriate in the correct circumstances and are often more valuable than large things. Both have their place."

Harmony exists where various parts come together as one. Let's take a soup, for example. Water, fire, meat, salt, etc., are all put together in one pot in order to bring out a complete harmony of taste. Also in singing, it is more appealing to hear a choir than a

Fig. 1.1 A portrait of Korean Buddhist scholar Wonhyo (617–686) housed in the National Museum of Modern and Contemporary Art in Korea

solo act. This is called the effect of harmony. Although from the Earth to the sun everything in between is said to become complete with the aid of the sky and the Earth, it cannot do so until it is harmonized by humanity. This meaning can be found in the harmony of tradition and newness. The philosophy of "Pungryu (taste for the art),"[1] a Korean philosophy, lies in the coexistence and harmony within nature, art, humanity, and all other living substances. The Cittaslow movement aims to seek to improve one's quality of life, as well as economic prosperity in harmony.

The secret of life is hidden in the "slow" and "curvy" that only God knows. If one wants to take the real approach to life, he/she is advised to enjoy life "piano, piano," the Italian phrase that means "slowly, slowly." If one lives slowly, he/she is able to enjoy a better life. By enjoying the benefits of "slow" every day, it brings a "higher quality of life" as well as a harmonious rhythm of life.

"No slow, no life." Where there is no "slowness," there is no life and no hope. If "fast" is regarded as straight, "slow" is a curve. In

[1] Pungryu is an elegant culture that was enjoyed by Koreans; they visited beautiful mountains and valleys reciting poems and playing traditional Korean instruments since ancient times.

Chap. 22 of the book *The Scripture of the Way and Virtue*, the philosopher of ancient China, Laozi (sixth century BC), said that only if something is bent, it is complete [2]. Even the poet Jung-Hee Moon, the poet of Korea (Republic of Korea), agrees that only if something grows naturally, whether it is out of shape or beautifully bent, it is the essence of completion; that is the natural beauty of life. Unfortunately although a human being is one of the most beautiful curves created by God, it lives in a built environment full of straight lines such as expressing a heartless world. A city is entangled in straight lines, while nature is curvy. Fast is straight, but slow is curvy and patient. A city is a product of the human will, but the countryside follows the provision of nature.

The river, mountains, sun, moon, seashore, and fruits are curvy. Rain that drops from the sky down straightly flows over the Earth and swirls in curvy stream. A straight line seems impatient and cold-hearted, while curvy is composed, warmhearted, and relaxing. The curvy is well represented in melodies and rhythm of music and a crooked tree. When all is said and done, the best curvy is the eyes of the human beings, and many artists represent them in their works, like "Ochi Chyornye" (Dark Eyes), a Russian folk song.

That is, life is not straight, sweeping everything before one. People who are not mild or composed may cause troubles with others, because such mildness and composure comes from only the curvy, not the straight. It is time to learn the slowness, mildness, and composure in a life and how to love from the curvy. As people often face many repeated complications in life, in the end, they appreciate the beauty of the curvy. The reason why love is the curvy is that it embraces everything in the world. Love and art come from the harmony of the straight and the curvy and serve as a driving force to people's lives. Such power of the curvy can be seen everywhere, fishing lines in the movie, "A River Runs Through It" (see Fig. 1.2),[2] an icon of our times, the Beetle (see Fig. 1.3) of absolute sense, and newly marketed curved TVs.

There is a craftsman who has earned a living by making mandolins for 30 years in Sorrento, Italy, where everything can be enjoyed under the sun. The master craftsman says that the tone and sound quality of

[2] "A River Runs Through It" is directed by Robert Redford in 1992.

Fig. 1.2 A poster of "A River Runs Through It" (*Source*: movie.naver.com)

Fig. 1.3 Volkswagen Beetle (*Source*: en.wikipedia.org)

a mandolin depend on its curvy shape. Such is a mandolin's secret, the eternal truth of the natural world has continued with a curve. Therefore, as shown in Fig. 1.4, the secret of life is hidden in "slow" and in a curve that only God knows. The city environment where people live today is well described in the lyrics of a song "Square's Dream" that was released in Korea by W.H.I.T.E. in 1996.

Fig. 1.4 There is a receptive royal tomb through the naturally formed pine trees. The city created by humans is straight, whereas the nature created by God is curvy. By courtesy of Chosun Daily

Square's Dream
sung by W.H.I.T.E. (1996)

Open my eyes in the square bed

The same scenery through the square window
Open the square door and sit on the square table

After read the square morning paper
Put the square books into the square backpack
Take the square bus and pass by the square building
Enter the square school and again the square classroom
The square board and the square desks
The square audio sets and the square computer, TV
The same day on the square calendar
I am only breathing without recognition.
(ellipsis)

Moses climbed up on top of the mountain and waited in order to be enlightened; on the seventh day, the Lord replied to him. Therefore, if one desires to hear the sound of God, waiting patiently is essential. Isaac Newton, the English physicist (1642–1727) [3], gave full play to his imagination while waiting under the apple tree; and Archimedes, the Greek mathematician (287–212 BC), did the same while lying in a bathtub. They were both having a big dream.

Even the classical Greek philosopher Plato (427–347 BC) was not involved with the so-called violent debates and discussion when he/she was meditating with the companions of Plato's Academy in the Garden of Academos. Looking at *The Dialogues of Plato*, they all refer to a very casual and freestyle of talking. "Slow" goes hand in hand with "waiting" and "leisure." During his 18-year exile, the scholar of the late Joseon Dynasty [4], Dasan Yak-Yong Jeong (1762–1836) who is likened to famous Renaissance man Leonardo da Vinci, left the undying legacy of more than 500 books through finding the beauty and happiness of "slowness."

Slowly but surely. Following the philosophy of "slow," Korea was the first to publish the Buddhist scripture "Jikji"[3] printed with the

[3] The Buddhist scripture printed with the world's first movable metal type developed in Korea during the Goryeo Dynasty is at least 78 years older than the first Gutenberg Bible.

world's first movable metal type, and the Duomo Cathedral in Milan, Italy, could be established that the American author Mark Twain described as "a poem in marble." Every masterpiece needs "slowness" and "calmness." The reason why Japan is so advanced in designing products starts with artisan spirit, whose masters work slowly but surely on every detail.

It tends to take a long time for a masterpiece to be completed. For example, in Barcelona, Spain, the Antoni Gaudí's unfinished masterpiece "la Sagrada Família (Basilica and Expiatory Church of the Holy Family)," begun in 1882, is still under construction, and it will take at least another decade for completion. Korea possesses a total of 10 UNESCO world heritage sites and has the third most UNESCO world heritage sites in Asia after China and Japan as of 2014 [5]. The world's first metal publication "Jikji" (see Fig. 1.5) was published in 1377, which is 78 years earlier than Johannes Gutenberg's

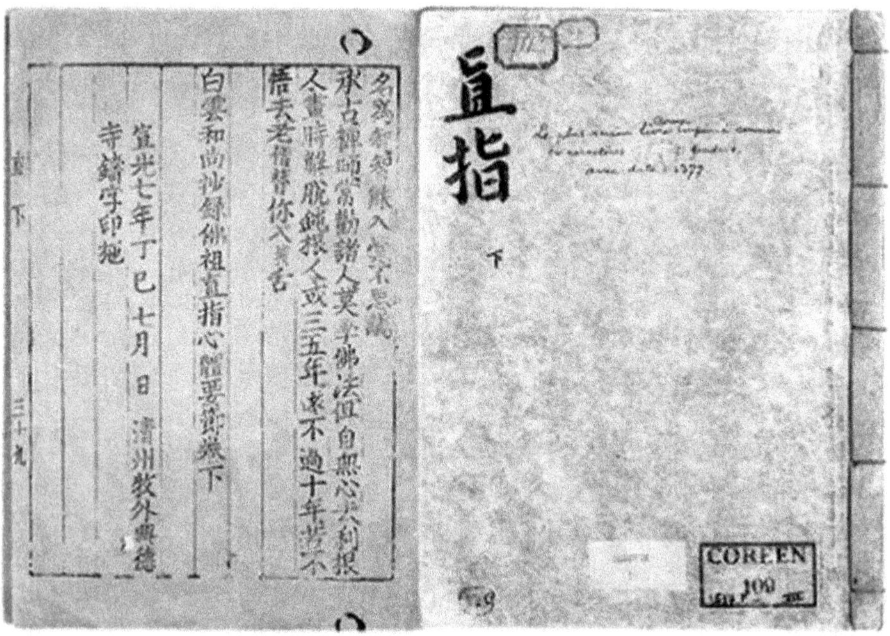

Fig. 1.5 Printing is humanity's biggest contribution to the communication media. The world's first metal publication "Jikji" was published in Korea in 1377 prior to Gutenberg's metal prints in 1455. The second volume has been preserved at the National Library of France (*Source*: commons.wikimedia.org)

Fig. 1.6 "Tripitaka Koreana" in Haein Temple located in Gyeongsangnam Province, Korea (*Source*: en.wikipedia.org)

"42-line Bible" printed in 1445. Furthermore, another Buddhist script *Mugujeonggwang Daedaranigyeong* (*Pure Light Dharani*[4] *Sutra*) is the oldest remaining book printed on a woodblock, estimated to have been published in 751. Robert Buswell at the UCLA Center for Buddhist Studies stated that the Korean collection of the Buddhist scriptures, "Tripitaka Koreana"[5] (see Fig. 1.6), is one of the world's greatest cultural achievements and is comparable to America's lunar exploration in the 1960s.

"La Sagrada Família" is one of Barcelona's most popular tourist attractions, and the construction of this church will continue for at least another decade. This church, a work on a grand scale, began in 1882 by the Spanish architect Francisco de Paula del Villar (1828–1901). At the end of 1883, Gaudí (1852–1926) was commissioned to carry on the work, a task which he/she did not abandon until his death in 1926. Since then different architects have continued the work after de Paula del Villar's original idea. After Gaudí's death in 1926, construction slowed dramatically due to a lack of funds and the outbreak of the Civil War (1936–1939). The construction pace started to pick up again in the mid-1950s, and now two façades and eight towers have been completed. The main nave was roofed in 2000. At that time construction was expected to last for another 100 years, but modern technology has enabled architects to speed up construction so that the church is now due for completion before 2030. The

[4] A mystic Buddhist formula

[5] The complete collection of Buddhist scriptures carved on over eighty thousand woodblocks

progress has been slow, but the result is already considered to be a masterpiece.

During Korea's Goryeo Dynasty (918–1392), two collections of all the sacred writings of Buddhism were created to overcome the crisis of the nation. One of them is the Tripitaka Koreana (1251), which was carved onto 81,258 wooden printing blocks. It took a long time to complete the Tripitaka Koreana through repeating various processes. First, the wood of Korea's wild cherry trees was soaked in salt water for 3 years, then cut, and then boiled in salt water. Second, the blocks were dried in the shade for 3 years until they were finally ready to be carved. After being carved, each block was painted with poisonous lacquer to prevent insect damage. It took 2 or 3 weeks to carve the wooden blocks needed for one page, and the carving of the whole work took 16 years to complete. Like the great legacies to the word, "slow" gives birth to matureness and mastery.

1.2 The Era of Sensibility and Deceleration

"Fast" takes the value of every day. "Slow" is the root and origin of human beings. "Slow" should be able to turn growth into maturity and quantity of life into quality of life, heal the soul, and help spiritual growth in order to build an everlasting future.

It is said that Native American people used to ride on the wide prairie, stopping every once in awhile in order to let their soul catch up so as not to lose it. This was called the "Indian Time." Through this way, we can see that "slow" is necessary in order to fulfill the soul. Even among the Arabs, there is a saying, "the soul travels at the speed of a camel." Thus, one cannot say that the statement "slow life is so beautiful" is a contradiction. By perseverance even the snail could reach the ark (see Fig. 1.7); accordingly there is a Chinese poem by Bai Juyi (772–846) [6], the famous poet in the Tang Dynasty (618–907), an imperial dynasty of China.

Facing Wine
by Bai Juyi (772–846)

On top of a snail's horn, what's there to squabble over,
with this body no more enduring than sparks from a flint?
Come riches, come poverty, be happy while you may –
the man you never see laughing – he's a fool for sure.

Fig. 1.7 A snail crawling slowly has many curves hidden in it. By courtesy of graphic designer Che-Ryung Sohn

There is a plaque of a short poem in the house where the wandering minstrel Kim Sat Gat (1807–1863) lived. Located in Yeongwol, one of Korea's slow cities, it says, "Everything in the world has an order, but only the life which is floating in the air is struggling." Even a poem of 200 years ago is criticizing the "fast life," and even the Bible says, "The fastest runner doesn't always win the race... (Ecclesiastes 9:11)." Why are people always rushing?

There is a report stating that people today live in average 40 times faster than in the past [7]. However, each tempo has its own intrinsic meaning. The leopard and the gazelle living in Africa depend on their speed in order to survive. In contrast, the sloth in South America has to carefully blend in with the trees and moves only slowly and thus does not attract attention in order to retain its life.

Thus, there is a natural speed for humans. However, since the Industrial Revolution, our lives depend on our work, and therefore, humans tend to have lost control over time. Today's people suffer due to the speed of life and are caught within the excitement and compulsion; hence, they are controlled by time from the moment they wake up. As everything seems to get faster and the time to concentrate on every detail is shortened, humans lose the feeling of relaxation. Italians have a saying, "man measures time and time measures man" [8]. It is necessary to understand that time is not meant to be used urgently, but is one part of a long and slow process.

Many people today are living life as a stone that is thrown away. Almost all people want to be at the top within the blink of an eye, while hastening to reach it. If there should be but one person who would prevent them from reaching it, they would get angry and shout out in anger.

In the book *Der Papalagi* (see Fig. 1.8) published by Erich Scheurmann in Germany in 1920, the speech of a Samoan chief named Tuiavii therein gives us a deep impression. "We have to rescue the poor and misled Papalagi (white man; a person who has ripped the heavens in the Samoan language) [9]. We have to get back his time. We have to teach him that the time from sunrise to sunset is so long that no person is able to make full use of it."

There is a term called "slow politics" [10]. Democracy is based on the power of the people; therefore, handling a job swiftly cannot be considered a democracy. It is necessary to invest more time into politics to create a better place for oneself and one's descendants. However, if there is no patience, it is impossible to establish a democracy. The effort of gathering people together to have a discussion, while listening to each and every opinion, respecting minorities,

Fig. 1.8 "Papalagi (white man)" is referred to as the person who has ripped the heavens in the Samoan language (*Source*: amazon.co.uk)

and finally coming into an agreement, is the meaning of democracy and thus is by nature "slow politics."

"Fast" takes the value of every day. On a Friday morning of January 2007 in Washington D.C., the subway exit was crowded by business people. Along came a youngster wearing jeans and a baseball cap holding a worn-out violin and looked at the crowd quite seriously while starting to play a tune. However, the busy people passing by did not seem to notice him at all. The 64th person passing by finally noticed the young violinist and turned his head toward him. After 6 min into his play, one person stopped and listened to his music while leaning against a wall, and within 43 min, seven people stopped for a little over 1 min to listen to the young violinist's concert. In the end, 27 people had dropped money into the violin case, and thus, he/she gathered a total of $37.17.

The next morning, the people were shocked reading the newspaper. The person, who had performed in front of the subway exit, was the world famous violinist Joshua Bell. He/she was holding a $3,500,000 Stradivarius and was performing for 45 min for free on the streets; however, 1,070 passing people were too busy to notice and could not even spare 1 s of their lives. The Washington Post that suggested this experiment regretted to announce that today's people are chased by everyday life that they do not notice the things that are happening in their immediate surroundings. Let us think, if we are not one of those who walk toward a destination while missing the value in the beauty of flowers on the sideway and the blue sky or the smile of people passing by.

Save your own soul. In 2010, the movie "Black Swan" (see Fig. 1.9) was screened and attracted the attention of the world. The main character of the movie had a hard time getting the main role in the ballet "Swan Lake" and finally managed to move the audience through her outstanding performance. However, in the synopsis the main character got entangled in the speed, competition, pressure, and other ugly human characteristics during the process and finally had her soul destroyed.

If the soul is not safe, it is impossible for humans to survive. The Apple Company is known for having more liquidity than the American government. Steve Jobs (1955–2011) had been considered the world's most successful CEO, being the leader of the speed by which the Apple

Fig. 1.9 The movie "Black Swan"; the main character struggled to be a prima donna and then finally had her soul destroyed (*Source*: movie.naver.com)

Empire had been growing. When the iPad 2 was introduced, he/she claimed that the speed had increased unbelievably, and while giving a performance, he/she was welcomed by a storm of applause by the participants. On the 40-year anniversary of the publication *Future Shock*, the American writer and futurist Alvin Toffler also stated, "the world will become even faster and busier in the future."

"Cittaslow" is the combination of "slow" and the "city." In every city, there are two factors that symbolize speed. One is the car; the other is digital media. The car keeps on increasing in speed and quantity. A city where the car has priority over people is abnormal and makes us unhappy. As shown in Fig. 1.10, the speed of the turbo capitalism[6] is so fast that it causes damage to the capitalist system. It is as "the world changed from big things eating small things to fast things eating slow things" [11]. In the world of capitalism, "survival of the fittest" is the law that starts to dominate.

[6] Turbo capitalism refers to an unregulated form of capitalism with financial deregulation, privatization, and lower tax on high earners.

Fig. 1.10 The World Trade Center (*left*) attacks on September 11, 2001, and Korea's national treasure No. 1, "Sungnyemun" (*right*) arson in 2008, are a warning to people who move with speed not paying attention to their surrounding while getting into conflicts and seeing the disappearance of righteousness (*Source*: *left*, 2002 Pulitzer Prize, Spot News Photography, Steve Ludlum, New York Times/right, by courtesy of Chosun Magazine)

If one is going too fast, no matter if saint or artist, he/she is doomed to fail. Since they are striving to succeed quickly, they miss valuable opportunities. "I am busy, therefore I am" is the statement that drives the Korean society into states of anxiety and depression, conflict, and competition and causes the disappearance of righteousness through corruption.

The global pharmaceutical company Eli Lilly and Co. recently surveyed men and women aged 34 from 13 different countries about their sexual life pattern. Portuguese and Mexicans were the most sexually active, with couples engaging in sexual intercourse 2.05 times and 2.03 times per week, respectively. The survey also showed low frequency rates in Finland and Denmark where the results stood at 1.21 times and 1.10 times on average per week, followed by Korea with an average of 1.04 times per week, only half of Portugal.

A snail has its own speed. Just as a snail going full speed at all times cannot be considered a snail, a person going full speed at all times is hardly human. Even so, the word "slow" is still considered an insult [11]. In the Oxford English Dictionary, "slow" is defined as late, dense, boring, lengthy, draggy, and sluggish; thus, it is not surprising that humans naturally connect the word with a negative image.

1.3 Smallness: To Metrion Life

Small things last forever. "Small" includes everything that is moderate, appropriate, nice, and beautiful. Plato's "to metrion (due measure)" is a keyword in his "Late Dialogues." In Philebus he/she has shown that both the good state in the cosmos and the good life in human beings are completed. In his "Statesman," he/she argued that all the good and fine things are produced by preserving the metrion. The abnormal demand to limitlessly strive for "faster," "bigger," and "stronger" in today's society is encroaching one's freedom like the spread of a cancer cell. Therefore, it is necessary to recognize the value of slow and small things.

In the book *Small Is Beautiful*, the socialist and economist Schumacher (1911–1977) describes small things as free, creative, meaningful, comfortable, happy, and everlasting [12]. That is, it can be said that we can have fun and enjoy our freedom to the fullest if we "stay small." If one wants to maintain certain economies of scale, it is essential to establish a pleasant environment first, where humans' happiness can coexist and the human being is in the center of importance. The beauty of small things can therefore be defined as the beauty of the human scale [13]. Only in a suitable scale, it is possible to gain a "tender loving care," which would be not obtainable without the optimal scale. According to the Bible, "... You have been faithful with a few things; I will put you in charge of many things... (Matthew 25:21)," and "The LORD did not set his love on you, nor choose you, because you were more in number than any people; for you were the fewest of all people (Deuteronomy 7:7)."

The industrial and technological progress of an enterprise is based on the logic of "economy of scale," and in such an environment, human relationships are put secondary. In a big school, students are given a number, just as the patients in a big hospital, or the employees in a big factory.

Only in a small group people are able to find their own identity and demonstrate their individual diversity. That being so, it is essential that even in big organizations, small groups are established. Thereby big organizations can keep the rule and order by their centralization, while seeking creative freedom and disorder by decentralization. However, this diversity and freedom is what brings great creations. The most significant value of Louis Vuitton over the last 150 years has been attention to not only the visual details but to the details in storytelling which promote its products' value. The business model of small business, for example Starbucks' is based on the formula "everything matters," thus paying great attention to detail when serving every single cup of coffee.

Happiness comes from minor considerations. Schumacher said, "small is free, efficient, creative, enjoyable, enduring." Originally, happiness has always been something small rather than big. The Korean novelist Do-Eon Kim also indicated that consideration is the key to share minor happiness. There was a youngster waiting at the bus station for the bus, and he/she was about to light a cigarette. Just before doing so, he/she noticed a pregnant woman who was passing by, and he/she waited until the woman was out of reach before lighting his cigarette. Such consideration is what brings minor happiness.

In the collection of essays "Über das Glück (Happiness) (see Fig. 1.11)" [14], Hermann Hesse (1877–1962), a German-Swiss poet, novelist, and painter, indicated that the small leisure of our lives has already been taken away. However, he/she introduced his

Fig. 1.11 *Über das Glück* (*left*) by Hermann Hesse says that moderate enjoyment is double enjoyment – and do not overlook the little joys; "Mortals and Others" (*right*) by Bertrand Russell states that the best person is pleasant, bright, and tender (*Source*: *left*, amazon.de/*right*, amazon.com)

philosophy to enjoy life: moderate enjoyment is double enjoyment – and do not overlook the little joys.

The British philosopher Bertrand Russell (1872–1970), who was awarded the Nobel Prize in Literature, also confirmed "the best person is pleasant, bright and tender" [15] (see Fig. 1.11).

Hesse and Russell were both people who knew how to enjoy the small things in life. Actually the whole world is smiling at us; it is just that we are too shallow to see it. Even in the poem "The Power of Laughing" [16] by the Korean poet Chil-Hwan Ban, we can find the happiness and enjoyment from minor things in the daily life.

The Power of Laughing
by Chil-Hwan Ban

A wild rose is climbing over a wall
You can be caught in the very act

It just smiles at us.
All smile along with it, not catch it.
(ellipsis)

Young-Mo Ryu, a social thinker of Korea (1890–1981), described the Korean word of a face as "window of spirit." The more frequent people laugh, the more traces they have on their face.

The domination of "materialism and jumboism," which was the foundation of civilization, is long over. The author of the book *Microtrends*, Mark Penn, states that the microtrends have primacy over the megatrends and that megatrends were lived in the 1980s [17]. The proof lies in the vague boundaries between mainstream and nonmainstream, classic and avant-garde, and high and street culture. Rather than eating and wearing the same things, this time dictates individualism, and therefore, it is vital to concentrate not only on the majority groups but more importantly on the minor groups in society. The American changemaker Seth Godin also underlines the importance of "small" in his book *Small Is the New Big* [18].

The so-called power scale from the past has now rather fallen into a burden. Whether individuals or groups, if one wants to adjust to the fast changing world, it is more favorable to be small. In the twenty-first century, it is required to be small in order to survive. The human being is naturally a small organism and thus is beautiful and peaceful when he/she is small. Especially in the high-industrialized society, being "small is so beautiful, delightful, and peaceful."

In 2011, a newspaper company in Fujian, China, conducted an online poll on the word of the year and the Chinese word "Wei (small)" was ranked the number one. The reason of "Wei" being voted was that people came to realize the great importance of ordinary people and things. More recently, even in the entertainment industry, more and more people are moved by the small Bollywood[7] movies from India rather than the enormous capital investing blockbusters. Although people are not that different from each other, small changes make a big difference.

Let's have a look at an example to find the reason why small things are good. Any seed or egg is small. A priest of a church once said that first there must be a small stream in order to become a big river and eventually to become an ocean. Furthermore, he/she added it would be necessary to expand the amount of small churches in order to establish a good religious foundation and finally to be able to build a healthy mother church. Jesus said, "It is like a grain of mustard seed, which, when it is sown in the earth, is less than all the seeds that be in the earth: But when it is sown, it groweth up, and becometh greater than all herbs, and shooteth out great branches; so that the fowls of the air may lodge under the shadow of it" (King James Bible, Mark 4: 30–32).

The same goes for a big car and an apartment. Lately, even in the housing market, small places are becoming the trend. The Korean Ministry of Land, Transport and Maritime Affairs sets the minimal living space per person to 151^2 ft (14^2 m), which is exactly the same as the cabin that the American author Henry David Thoreau (1817–1862), best known for his book *Walden*, lived in near Walden Pond [19].

In traditional Korean houses, units of space per person determine the size of the building. Taking four pillars and leaving one unit of space in width and connecting them, it becomes a room, which will most likely be smaller than 86^2 ft (8^2 m). The prominent Korean Confucian scholar, Yi Hwang (1501–1570), stacked over 10,000 books in his house of 4.5 units running to about 388^2 ft (36^2 m).

[7] Bollywood is the informal term popularly used for the Hindi-language film industry based in Mumbai (formerly known as Bombay), Maharashtra, India (*Source*: http://en.wikipedia.org).

Some Eastern famous scholars lived in such a tiny room or house but enjoyed the nature surrounding it, gathering knowledge and understanding the ways of the world.

The author of the book *The Small House Book*, Jay Shafer, saw the collapsing housing security due to the financial crisis in the United States, resulting in many people not being able to pay back their loan and falling into debt. Consequently, he/she decided to live in a small house with only the necessities rather than living on the streets and hence built the small trailer "Tiny House" barely 89^2 ft (8.3^2 m) in area. As a result of this initial step, there was a huge boom into the market for "Tiny House." The reason why many children nowadays are more interested in a microscope rather than a telescope is also due to the fact to find something big in small things.

1.4 Nowadays, Small Means Big

A solution to live a small life. The Volkswagen Beetle from Germany, which takes its form after the bug, is a rather cute compact car that was first launched in 1938 and shortly after was overwhelmed by popularity among the young feminine market. During the 1950s, the American market was filled with big "gas-guzzling" cruisers. However, with just one copy of the advertisement "Think Small," Volkswagen Beetle boomed the market. This advertisement was for all those people who were deceived by thinking only big things have good value, and it gave a fresh alternative substance. Today, even 80 years after its launch, the beetle is still considered the top brand among compact cars (see Fig. 1.12).

The publisher of the global affairs magazine "Monocle," Tyler Brûlé, once wrote a weekly column entitled "The secret of my brilliant Korea" in the "Financial Times" magazine on January 7, 2011, and listed top 10 Korean cultural products that are likely to succeed overseas, including Incheon International Airport, department stores, and Bindaetteok or traditional green bean pancakes. In an interview with a Korean newspaper, he/she suggested some improvement points in order to become a cultural advanced country. The first one was the "bookstore." When he/she was waiting at Gimpo International Airport in Korea, he/she stopped at the 7-Eleven convenience store to buy a magazine; however, he/she could not

Fig. 1.12 Korea's neighborhood electric vehicle (low-speed vehicle); the new Beetle opened the market for compact cars, and the trend will shift to slow and clean electric cars in the future. By courtesy of ChosunMedia

find anything as most books or magazines were not sold in the store anymore. Even in the university area, it is hard to find a bookstore nowadays. The second was the "cultural intimacy." Brûlé said that the former mayor of Seoul, capital city of Korea, Se-Hoon Oh's "Design Seoul" project was encouraging; however, planting trees and widening the bicycle roads were not enough to build a "design city" [20]. It is also important to open up small restaurants and shops, which make an area cozy and establish small spaces and alleys for people to take a walk through and to appreciate.

Seoul is so focused on developing the "new town" that the "old town" is completely forgotten and is slowly disappearing. To enhance the quality of the culture, more new generation entrepreneurs need to not only make money but also enjoy easy living. In many developed countries including Japan, for example, more and more of the younger generation who are looking for work are not interested anymore in competing for a job of a major company, but rather prefer to run their own steak restaurant with the best steak while enjoying their life. Brûlé's statement confirms that the concept of "Cittaslow" is going into the direction of an advanced country.

Gimhae City in Korea differentiates itself by establishing the brand image "Gimhae Reads" and has created the "Mini Library" in township offices, restrooms of a park, a jail, etc., and the "Very Small Library" at bus stops in order to fill the city with books and encourage a reading culture. Also, Gimhae is soon to construct a multicultural library. In the United States, many cities such as "One City One Story" of Boston, Massachusetts, are operating "One City One Book," a generic name for a community reading program that attempts to get everyone in a city to read and discuss the same book. Today, One Book programs are held in libraries, colleges, bookstores, and cultural centers across North America, Australia, and the United Kingdom. In Dublin, Ireland, for example, there is "Dublin: One City, One Book" that is an award-winning Dublin City Council initiative which encourages everyone to read a book connected with the capital city during the month of April every year.

Among the businesses in Kyoto, Japan, there are 33 commandments in order to establish a successful business. One of those states, "Don't operate two different businesses." Such Kyoto businessmen are also called the "1,000-year businessmen." There are a total of six businesses that have been operating for more than 1,000 years and more than 1,600 businesses that are 200 years and older. Such businesses are often called "sustainable businesses" that have been doing the same thing for generations. When asked what their family motto is, they all agree to one statement, "Study day and night."

In the Nishiki Market called "the Kitchen of Kyoto," there is Iyomata, a sushi restaurant that opened in 1617 [21]. This restaurant has four small tables in one little room and serves Saba sushi (mackerel sushi) as its main dish. The number of customers is 60–70 on an average a day: however, more than 100 people take away mackerel sushi every day. Despite not opening any franchises, it has been thriving for more than 400 years, and it makes sure that every sushi meets the customer expectations and does not disappoint.

Big organizations make people small. People have moved from farmland and small towns to the big cities, and this led to the abnormal growth of "megalopolises." This "gigantism" of the huge cities followed by the thought that all good things are big has had a negative effect on people's psychology and social structure. If there are too many people stuck in big businesses or big governments, they tend to make the people feel smaller inside in the presence of a big structure and become like machines.

Despite the information age, governments currently have too many things and people to take care of. However, it is still unknown whether a big government is the right solution for economic and financial matters. In the 2012 US presidential election, voters agreed that they did not desire a big government any longer. They rather demanded a small government with an understated budget that is based on liberal principles. It is an urgent matter to think about what a government is actually supposed to fulfill and rather to readjust and change to a smaller government and even bigger citizens.

Small countries live well and the trend goes toward small- to medium-sized businesses worldwide. The richest countries in the world include Switzerland, Austria, Netherlands, and Singapore, which have a small land area in common. For example, Switzerland has less than eight million inhabitants [22]; nevertheless, it is divided into 20 cantons.[8] Each of the cantons is considered a development unit, which has different cultures and independent economies.

Korea is also a small territory; however, it is a small but strong country that has half succeeded through a "small country, but big business" strategy. After the Korean War (1950–1953), Korea was at the same economic level as Africa, but according the World Bank, in 2003 the country held the 11th rank among 196 countries in the world, and its GNI (gross national income) per capita surpassed the USD 20,000 mark for the first time in 2007. Korea is already more advanced than other developed countries in fields of cars, ships, airplanes, steel, computer chips, cell phones, and home appliances. Korea is economically healthy considering its small size, and therefore, now is the time to find good living/well-being/the way to live happily.

1.5 What Is "Cittaslow^Slowcity"?

Cittaslow, the Italian for Slowcity, strides its grand step in Greve in Chianti, one of the small towns in Italy back in 1999. The idea is to develop a slow town, which means slow concept is well spilled over in development project, and it is a very opposite way to those of big

[8] Canton is a political subdivision in France, Switzerland, and other European countries (Britannica Online).

city. The happy community building project respects legacies of blessed nature and cultural heritage and requires local residents to involve voluntarily. Their sustainable goal is the revitalization of community: to moderate negative impact on environment, to respect Mother Nature, and to enjoy culinary tradition and unique charm of culture.

Why are we embracing the Cittaslow movement? If asked this question, the answer is "simply to live like a decent human being." People have sacrificed the delight and happiness of "slow" for the convenience of "fast" like ecstasy presented by the current revolution of objects. Those advantages coming from speed should be put into our hands in order to enjoy life and create happiness. The philosophy of the movement is to tilt the weight to matureness instead of growth, life quality instead of quantity, and respecting depth and dignity instead of speed. The "slowware" should consist of "slow," "small," and "sustainable" (3S).

The Cittaslow movement does not aim to turn the majority of people from a "fast-based lifestyle" to a "slow-based lifestyle." "Fast" can be thrilling, productive, and powerful. If it were not for the "fast" lifestyle, some countries like Hong Kong, Korea, etc., would have stayed poor until today. The problem hereby is not to reverse the time, but to find a harmonious rhythm of life between fast and slow, quantity of life and quality of life, rural community and city, local and global, and analog and digital. The movement is the recipe to find the balance between "la dolce vita (the sweet life)" and the dynamic information society.

Although the Cittaslow movement seems unrealistic in the stressful society, as of June 2015, 200 member towns in 30 countries are already participating in the movement, including eleven Korean member towns. The Cittaslow Manifesto is shown in Table 1.1.

> ...Cities given life by people with a respect for times past, rich in places to meet, in theatres, workshops, cafes, restaurants, spaces for imagination, with intact landscape and skilled craftsmen and handicraftsmen, where people can still appreciate the slow wholesome passing of the seasons, with good eating and good health provided by natural foods, with respect for tradition... Greve in Chianit, October 1999

Table 1.1 Cittaslow International Network: 200 cities in 30 countries (as of June 2015)

Europe (18)	Oceania (2)	North America (2)	South America (1)	Asia (6)	Africa (1)
Italy (77)	Australia (3)	Canada (2)	Columbia (1)	Corea (11)	South Africa (1)
Austria (3)	New Zealand (1)	USA (3)		Turkey (10)	
Belgium (6)				Turkish Republic of Northern Cyprus (1)	
Denmark (2)				China (3)	
Finland (1)				Japan (1)	
France (8)				Taiwan (1)	
Germany (12)					
Great Britain (5)					
Holland (7)					
Hungary (1)					
Iceland (1)					
Ireland (1)					
Norway (4)					
Poland (20)					
Portugal (6)					
Spain (6)					
Sweden (1)					
Switzerland (1)					

Cittaslow—international network of cities where living is easy (see Fig. 1.13). No matter how successful a human being is, he/she is always dependent on nature and is living just like riding on the back of a snail. A place where humans and snails can live together is a true community and a place full of happiness. Snails essentially cure themselves and try to survive their whole life long while continuously moving toward a destination [23]. This idea can be transferred to a community that needs to cure itself to survive and continuously strive toward a communal goal. Just as the organs of the snail are placed within the snail house, the logo of the Cittaslow movement is also a snail carrying a city on top of its house to symbolize the sense of community among humans. The motto of Cittaslow International is a storytelling of the international network of cities where living is easy. The aspect that humans should concentrate on is to create a good life and establish a good society. The quality of life isn't easy. It never has been, and it never will be. Nevertheless, we should do our best to accomplish it.

Cittaslow happiness. Korea, one of the poorest countries in the world after the 1950 Korean War, had focused on economic efficiency and increasing productivity to overcome poverty. Accordingly

Fig. 1.13 Snail, the logomark of Cittaslow^{Slowcity}. By courtesy of Cittaslow International

while the nation has had an unprecedented 60-year period of economic development, the people have missed something valuable. The slow city movement in Korea aims to observe nature, tradition, and community. These are fundamental for sustainable development in human society. Good for nature is good for humans. Human nature means true nature of the human being and saving nature is to save humans. In the big splash of globalization, local traditional identities are crumbled away. The Cittaslow movement provided small local towns with economic power, the local-colored, differentiated, and high-end brands, which empower them to face up to big countries or big cities' mass production. Koreans traditionally have honored harmony with nature, art, and human nature, and such community harmony will increase where there is economic coexistence, excess of mirth, and happiness and full-blossomed joy with the Cittaslow members in the world.

1.6 Points

If fast is regarded as straight, slow is curvy. The secret of life is hidden in "slow" and in a curve that only God knows. The Cittaslow movement endorses "a bit slower," "a bit smaller," and "a bit richer" instead of "more," "faster," and "wealthier" and builds on the 3S (slow, small, sustainable) in order to ensure permanence. If you live a little slower, the world becomes more beautiful.

Slow and happiness are the root and home of human beings. Through "Manbonism" we should embrace matureness instead of growth, life quality instead of quantity, and the depth and dignity instead of speed in order to prepare for a long and sustainable future. Through speed, we have lost the essentials since the Industrial Revolution and were controlled by it, instead of taking control of our lives.

Small is little and suitable, brings peace to the soul, and is human sized through which humans find freedom and happiness. Having a small and simple wedding, to which only the immediate family and very close friends are invited, makes some newly married couples feel happier. A small shopping district with many mini shops and a retail district packed with specialized businesses bring individuality, culture, romance, and joy. Now, small means big, small countries are rich, and small and medium businesses are the general trend around the globe. In the twenty-first century, it is essential to redefine the value of "slow" and "small."

Why are we embracing the Cittaslow movement? The philosophy and skill of "Cittaslow," the Cittaslow International, and the symbolic meaning of the Cittaslow motto were explored. The growth pattern in modern society needs to return to the old-fashioned growth; therefore, the slow-moving Cittaslow network can be an alternative to the sustainable growth option and to the disaster of global ecosystems and degraded humanity.

The Era of Sensibility and the Delicacy of a Slow Life 2

Contents

2.1 Slow Life Versus Fast Life

"Slow" and "lazy" are completely different. "Slow" does not necessarily mean that one is not doing one's work. Take your time to do one thing efficiently and well. Work can be done in a much more creative way and human relationships can be improved, raising the overall quality of time. In the 250 years following the Industrial Revolution, everything has accelerated. Of course it brought a lot of advantages; however, it becomes obvious now that it brought many more disadvantages. Therefore, the "slow" movement within the global village is accelerating with tremendous speed.

According to the author of the book *Slow is Beautiful*, Carl Honoré, even in American cemeteries nowadays one can find drive-through [24]. Apparently people are supposed to drive through the funeral, while throwing the flowers out of the window in a time span of 90 s in order to remember the dead. In London, there is a gym where

d. sohn et al., *Go Slow and Curvy*,
DOI 10.1007/978-3-319-14854-0_2

"speed yoga" is offered, which crams the full hour of routine yoga into 20 minutes.

The digital world is getting busier every day. The purpose itself is to become busier in order to survive in the market. Smartphones are considered advanced guards of the digital era. Some people even call smartphones the "Jesus phone" [25]. According to a survey about cell phone usage, around 20 % of people pick up their phones while having sex and would speed up the process of sexual intercourse in order to get back to the phone. Such refers to the fast culture even entering the bedrooms of people.

People addicted to the digital life. Once you enter the digital network, it is difficult to live freely. According to a survey, the impulse to use digital networks keeps on repeating every 10 minutes. This means one lives only for the screen, through the screen, and with the screen. Registering on an SNS (social network service) makes it even harder to take one's eyes away from the screen.

Digital immersion is the enemy of depth [25]. Once one falls into the digital network, one does not move through one's own will, but according to the direction of others, and thus, one cannot hear one's inner voice. If one's own depth disappears, the depth and dignity of society also fade away. Youngsters in some European countries as well as in Japan and Korea with the highest Internet penetration rate are becoming addicted to online games. Although such games can be played for fun, as well as for educational purposes, the majority of online games waste one's daily productivity and diminish leisure time and freedom of one's mind.

If using the Internet or smartphones too much, the brain turns into a so-called popcorn brain, which refers to a brain that is not able to distinguish between reality and fiction. IBM is following a movement called "Slow E-Mail Movement" and refers to the checking of e-mails only twice per day; the British Prime Minister David Cameron on the other hand switches off his cell phone every time when participating in a meeting. According to the IT expert Nicholas Carr in his work "Not thinking humans," the English naturalist Charles Darwin (1809–1882) stated in his "Evolution Theory" that in the nineteenth century, the only organisms which survived were those that could adjust to the changes in the natural environment; however, in the twenty-first century, people have to hold back on the adjustment to the IT changes and have to control their digital consumption in order to evolve wisely.

Naturally, there is a positive side to it as well. On August 5, 2010, the incident of the mine disaster that occurred in Chile spread all across the world. During the incident miners were trapped 700 m (2,300 ft.) underground and passed their desperate struggle for life until being rescued on October 13, 2010. It was watching soccer and movies on the cell phones that helped them survive mentally.

A slow sensibility and thoughtful life. Now is the time to choose a slow life. The choice of how much time smartphones and digital media are going to consume our lives is up to every individual and is furthermore a philosophical matter. Let's try to go to a coffee shop in order to meet people rather than contact them online, to work out in a gym without screens, or to have one day in a living room without a television or any Internet-capable device. Let's enjoy a life where on weekend evenings, we turn to our families when the modem stops, look for nature, or meet the neighbor on the street and joyfully dance to bring the good energy into our lives. Let's cook, knit, repair our bicycle, or use our bodies to move and dig out old tools and use them. Many philosophers have agreed that the true and happy life lies in the past.

Jesus said that rich people couldn't go to heaven, which is due to wealthy people always being busy. The sin lies within their busy life. In order to live a new and happy life, it is necessary to take the time to feel and think slowly. This is due to the fact that if one is busy, the depth of sensibilities, the depth of human relationships, and the depth in our work will inevitably diminish. The poem "The Bridge" by the Korean poet Sung-Sun Lee shows the beauty and freedom of "slow."

The Bridge
by Sung-Sun Lee

I see a person crossing the bridge

(ellipsis)

After awhile there is another person crossing the bridge
Passing so fast that in a moment he lives apart from himself
After he has gone, only the bridge is left by itself

I see that people who cross the bridge fast make the bridge lonely

Ten rules for a healthy mind

1. See life positively.
   It will make you realize that there are always two sides to a coin.
2. Live with a thankful heart.
   It will give you energy in your life.
3. Greet people with delight.
   It will fill your heart with warmth and build a base for success.
4. Eat three meals per day slowly.
   It is the base of health, and gives birth to happiness.
5. Think about the other person's perspective.
   It will diminish your anger.
6. Praise anyone.
   The more you praise people, the more confident you will become, and praises will come back to you.
7. Be early for your appointment and wait.
   You will not be under pressure due to lateness and will build faith.
8. Smile, even if it's forced.
   With only a smile, it will brighten your mood.
9. Live honestly according to the law.
   Lying will only give you a sense of guilt and make you nervous around others.
10. Sometimes you need to miss out on something.
    It will put your mind at ease and someday you will receive something bigger.

Source: Korean Neuropsychiatric Association

2.2 Digital Addiction: The Enemy of Depth

If the Earth runs according to Earth's time, nature runs according to the time of nature. However, we hear that some vegetables and animals are grown much faster than natural time these days. For example, lettuce grows four times as fast and a small piglet takes 6 months to grow up instead of 6 years.

Everything in this world has its own structure. If this structure is shaken, there is chaos. However, due to today's mass transportation and speedy communication, the structure in many things is becoming more complex. This world we are currently living in is comparable with a big, unstable cargo ship. In our surroundings, many structures, such as the structure of society and psychology, have already collapsed; however, the bigger problem is that no one knows how to stop this phenomenon.

The cities where humanity disappears. The proportion of single households is the highest in developed countries. A special report of the Euromonitor International stated that Sweden had the highest proportion of single-person households at 47.1 % of total households followed by Denmark (40.0 %) and Norway (39.6 %) in 2011. The number of single-person households was 277 million or 14.9 % of total households in 2011 and is expected to rise to 331 million or 15.7 % of total households by 2020 [26]. However, "Doing Better for Families" published by OECD (2010) states that the average household size is smaller than 2.1 in Denmark, Finland, Germany, Norway, and the United Kingdom. Europe is undergoing significant demographic and societal changes. The population is aging; families are becoming increasingly diverse. Statistics from the European Commission reported that older persons (65 or more years old) made up 17.5 % of the EU-27's population in 2011 [27]. According to the US Census Bureau (2011), 27.6 % of all households consist of just one person—the highest in US history.

According to the Korean National Statistical Office (2010), the percentage of single households exploded from 4.8 % in 1980 to 23.9 % in 2010, and the family structure has changed from 4 head to 1–2 head families [28]. The number of senior citizens who live alone also rose from 548,000 in 2005 to 793,000 in 2010 by 45 %. This means that the overall population is aging drastically and the amount of potential working people is disappearing. This kind of unbalanced lifestyle and pattern is of deepest concern to the future of humanity. Therefore, a new lifestyle, such as "slow life," is needed in order to secure the sustainable existence of human kind.

According to the book *Going Solo: The Extraordinary Rise and Surprising Appeal of Living Alone* by Eric Klinenberg, "going solo" goes along with the current trends in the digital era, and although "playing alone," people don't feel lonely. In the info-communication revolution such as the Internet and social network services, we face the huge wave, individualism revolution, which is slow, calm, and soft, but tremendously fundamental, and will bring about a new democracy. Accordingly, single-person households are becoming the norm in the twenty-first century (see Fig. 2.1).

Currently, the urbanization rate in the world amounts to 50 %, and in 2011 the highest rate was reported as 100 % in Singapore, Hong Kong, Macau, Bermuda, and Monaco, 93.5 % in Venezuela, 92.5 % in Argentina, 89.2 % in Chile, 84.6 % in Brazil, 78.1 % in Mexico, and 72.8 % in Malaysia [29]. In the last 50 years, after the industrialization in Korea, the importance of industry and commerce has risen dramatically, resulting in an excessive urbanization. The rate in Korea already exceeds 90 % [30]. Korea, being a member of the OECD, stands among the highest-ranked nations for minimal energy consumption per capita and utilizing current technology in peoples' life pattern as a base. According to the World Bank from 2008 to 2010, Korea's energy use per capita was in the upper ranks among the OECD members along with Iceland, Canada, and the United States [31].

However, in the cities the degree of humanization is slowly disappearing. It needs to be taken into consideration, whether the trees, animals, and people's health and happiness are being respected and whether a place is created where people can enjoy themselves in a creative and productive workplace or whether these are all deprived.

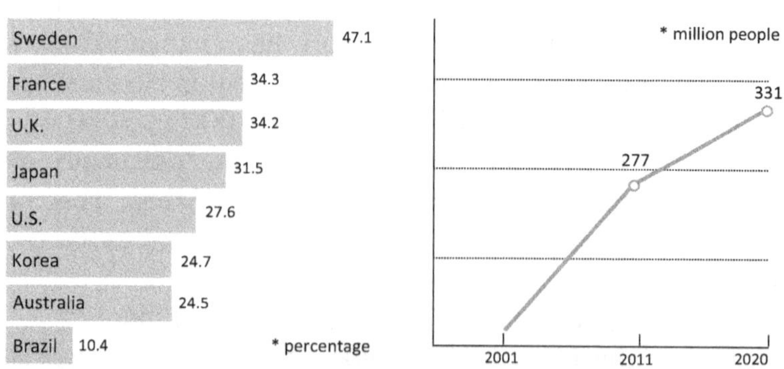

Fig. 2.1 Percentage of single-person households in 2011 and the 2020 outlook (*Source*: Euromonitor International)

Gradually, people in wealthier countries are increasingly realizing the cost of enjoying a "fast life."

There are many people who desire a new "creative culture." Life in the cities is technology based. However, technology should not substitute for humans, but should only be used in order to assist humans [12]. E. F. Schumacher said that he enjoys showing people who are visiting his home the manual-grinding mill. He bought the mill in order to grind wheat himself and to bake the bread that typically lasts up to one week to feed his family. On the contrary, he argues that the nutritive value of the food many people eat nowadays might be less than that of one sheet of paper. In general, humans who are using machines adjust to the rhythm of machinery, which is said to bring misfortune.

In the following dialogue, the relationship between humans and machines is pointed out. The pupil of the Chinese philosopher Confucius (551–479 BC),[1] Jagong, was walking along the road to the village when he saw an elderly man, who was using a jar to collect water and pour it on his field [32]. He asked the elder whether it wouldn't be faster to simply utilize a machine in order to water the field. The elder answered, "I heard from my teacher, using a machine makes you want to avoid work. If you avoid work, you will be thinking about a plot. If your heart is filled with plots, it will lose its pureness. If you lose your pureness, your mind and life will be insecure. If your life is insecure, you will lose the truth. It's not that I don't know how to use a machine, I'm just too ashamed to use one." Hearing this, Jagong ran away out of shame.

In Sweden, the government is trying to induce consumers to purchase products with less CO_2 emissions, while all groceries are labeled just like calorie points. For example, Max, the largest hamburger restaurant chain in Sweden, lists the carbon emissions next to every item on its menu (a large hamburger is responsible for 1.7 kg of carbon dioxide emissions). This shows that the government aims to reduce greenhouse gas emissions within the production process. On

---

[1] Confucius (551–479 BCE) was a Chinese teacher, editor, politician, and philosopher of the Spring and Autumn period of Chinese history. The philosophy of Confucius emphasized personal and governmental morality, correctness of social relationships, justice, and sincerity.

the other hand, the global village is trying to save energy with a special event called "Earth Hour," which encourages households and businesses to turn off their nonessential lights for 1 h. The voluntary energy-saving campaign is an enjoyable and romantic cultural movement, which lets people realize the beauty of darkness.

As Korea ranked eighth worldwide in terms of energy consumption in 2010 [33], there are many energy-saving movements being established. The Korean-based public enterprise High1 Resort has decided not to build a new building for their head office, but carefully selected an elementary school that was closed down previously and refurbished it in order to save the energy consumption used for construction.

However, global warming is having a significant impact on hundreds of plant and animal species around the world. Birds are laying eggs earlier than usual, plants are flowering earlier, and mammals are breaking hibernation sooner. According to a news release from Stanford University, if such ecological changes are now being detected when the globe has warmed by an estimated average of only 1 °F (0.6 °C), then over the next 100 years, many more far-reaching effects on species and the ecosystem will occur, when temperatures could increase by as much as 11 °F (6 °C) [34]. Springtime events such as blossoming, egg laying, and the end of hibernation now occur in an average of 5.1 days earlier per decade. The North American tree swallow offers a good example. Field biologists, who kept track of some 21,000 tree swallow nests in the United States and Canada over the last 40 years, concluded that the average egg-laying date for female swallows has advanced by nine days, a phenomenon that mirrors other North American studies confirming higher temperatures and the earlier arrival of spring.

Perhaps due to the climate change in Korea, the swallow in Seoul has already disappeared long ago. According to a Korean institute, in the past ten years, the population of swallows has decreased by 93 % [35]. Furthermore, according to a Korean university's research team, the average temperature in March and April in the past 40 years has risen in 53 locations by 1.5 °centigrade [36]. Global warming is shortening the winter preparations for all living organisms; if the temperature becomes warmer, flowers start to bloom, and insects start to move. While the life and activity of insects has become faster, the swallow arriving later in the year cannot find enough food to survive, making it harder to breed due to the changing life rhythm.

Once you are caught in the magic of speed and move faster, the pressure will always hunt you to move even faster. The ant is always floating through life; however, the spider is an expert in "stop and go" movement, sitting still and being able to abruptly run to catch its prey. The spider, moving just like a dance "slow slow, quick, quick," has a moving lifestyle people can learn from (see Fig. 2.2).

The Papalagi (the civilized man through the eye of Samoan natives) divided time into hours, minutes, and seconds and after it is shattered threw it away [9]. People live busily their whole life. People keep looking on the floor while walking, moving their arms back and forth in order to speed up their walk. Being infected by the speed virus, people enjoy the F1 Grand Prix, where monsters driving with a speed of 350 km/h pace through the track.

In order to change one's life from a fast to a slow pace, it requires a lot of effort. The vice president of the Korean financial company "Mirae Asset" Chang-Hee Kang says that his weaknesses are his fast pace while delivering a speech as well as while having a meal. However, in order to improve these, he holds a card stating "slow" in his pocket, which he looks at before every lecture. Additionally, not wearing a watch is another way to decrease the speed and become more relaxed toward time. The members of the Cittaslow club, "Jihe" that is run by the National Cittaslow Corea Network always cheer with the verse "slow slow, short short" in order not to drink moving from one bar to another.

"Cittaslow movement"—saving the health of the Earth. A warning regarding global warming has been published already. Mark Twain

Fig. 2.2 The spider, the expert of "stop and go," moving "slow, slow, quick, quick." By courtesy of ChosunMedia

said, “It’s not because we are not aware of it, but rather that we live with the vague idea that it’s not going to hit me and my immediate surroundings, which makes this topic so dangerous to encounter.” Whether it is a country, a business, or an individual, everyone has to take this issue into serious consideration. If possible, every family should plant as many trees as possible in their garden and show off about it. It would be right to say that forestry is the only measure in order to save our future and the health of this planet.

It is helpful to approach this matter with small actions, such as using a handkerchief instead of a tissue, to aid the planet. E. F. Schumacher, the author of *This I Believe: and other essays*, stated, “An ounce of practice is generally worth more than a ton of theory” [13]. It is important that especially in the twenty-first century, the 3R (reduce, recycle, reuse) movement is being considered.

Currently, people who want to escape the fierce competition for survival and the pressure of sticking to a rank in society are increasing. Instead they want a relaxed life and look for “downshifting.”[2] Their last choice is “slow and simple” [37].

Even China has joined the Cittaslow International movement and is spreading the new life of Cittaslow. In the global village, one can hear more good news (slow movement, humane world) rather than bad news, such as the era of materialism and money.

2.3 The Era and Culture of Sensibility

The most crucial emotion in Korea is the unique disposition of Koreans, “sinmyeong (ecstasy),” which is an exciting energy that combines different kinds of emotions such as joy, sorrow, pain, and hatred to desire [38]. Dangun, the legendary founder of Gojoseon, the first Korean kingdom (2333–108 BC), has passed down the ecstasy “sinmyeong,” the excellent combination of “pathos (passion), logos (logic), and ethos (desire, soul).”

[2] Downshifting is a social behavior or trend in which individuals live simpler lives to escape from the rat race of obsessive materialism and to reduce the “stress, overtime, and psychological expense that may accompany it.”

Full of the ecstasy "sinmyeong" and the taste for the arts "pungryu," Korea traditionally established an emotionally sensitive, bright, and clear national character. The Korean philosopher Seok-San Tak described in his work "How do Koreans live" that Koreans live with enjoyment. In enjoying life, they particularly like to eat and drink well and love to dress up.

In today's emotional times, the importance of sensibility management is increasing in the business environment. Although people experience sensibility in a rather intense way, the key to open sensibilities is culture. For example, the Korean musician Psy has become an "international sensation" through his song "Gangnam Style," and his music video has received over 1.7 billion hits on YouTube (see Fig. 2.3). "Gangnam Style" refers to a lifestyle in the Gangnam District of Seoul where people are trendy and hip and live a life of luxury. Therefore, Psy is singing sarcastically about the over-consumption and the excessive capitalism.

Leisure and freedom are the fuel of creativity. Standardization, division of labor, and assembly were the core values of "Management 1.0," which revolutionized the business world as the first management trend. However, in "Management 2.0," transformation, innovation, authority, and responsibility are the topics to deal with. The famous management strategist Gary Hamel announced, "The 20th century management schooling is dead" and stated that "Mass production as a basis for business as it was in the 20th century will not guide business in the 21st century" [39].

"Management 3.0" on the other hand values feeling, partnership, imagination, and creativity as the representative core values. The Canadian circus company "Cirque du Soleil" was founded in 1984,

Fig. 2.3 Korean singer Psy in the music video "Gangnam Style"

and by 2010, 100 million people had seen the show creating a profit of USD 929 million. Guy Laliberté, the cofounder and the current CEO of Cirque du Soleil, who is considered the king of circus, said, "we never reproduce our show." Furthermore, the artists of "Cirque du Soleil" come from 24 different countries and use "creativity" as the fuel to move on [40]. Additionally, he stated, "Inside every adult there's still a child that lingers. We're happiness merchants—giving people the opportunity to dream like children." Their motto is the quality of performance and respect for the audience.

The Associated Press asked four Nobel Prize laureates in an interview, "Where do you see the reason for an outcome?" They all answered that the point of success lies in the free choice of one's own area of study. Furthermore, the get-together with colleagues during breaks for a tea or a coffee and the time to talk casually are the base for creativity.

The same goes for us, as we also need leisure time in order to bring forth creativity. Three seconds are enough to find time to relax, and even when saying a prayer, one should wait for a couple of minutes in order to give God time to respond. The next "time for three seconds" is a humorous writing by anonymous.

Time for three seconds
Anonymous

Wait for three seconds before pushing 'Close' after entering an elevator.
You never know if there is someone who is rushing to get in.

If the traffic light turns green and the car in front of you is not moving, wait for three seconds and don't hunk the horn.
You never know if that person is experiencing a conflict that could change his life.

Wait for three seconds even if there is a car cutting into your line.
The driver's wife might be in an urgent situation.

When you part with your friend, wait for three seconds and watch him.
He might turn around and look at you, and you can give him a smile.

If you see someone on the road that just experienced a misfortune, close your eyes for three seconds and pray for him.
They will someday do the same for you.

Even if you are furious, look at the sky and wait for three seconds. Think about whether it is reasonable to get angry about such a trivial thing.

If you look out of the car window and see a child, who is looking at you wave at him for three seconds.
If he grows up, he will definitely do the same for your child.

If you see a person, who has sinned and goes to prison, do not curse him but think for three seconds.
If you were put in the same situation as that person, how would you have ended up?

If she is angry and keeps yelling at you like a waterfall, smile at her for three seconds.
You never know if she will laugh with you in the evening and have a glass of wine with you.

While at night, sensibility dominates, at daytime reasoning dominates. If you argue whether it is called night and day or day and night, we believe it is right to say night and day. The reason for that is, when people think about the order of sensibility and reasoning, sensibility should come first. The reason why people always claim that it has become harder to live is also because there is not enough time to rest and enjoy romance. Ten hours of a day are dark in average and we can see how important the darkness is for mankind. Just by looking at the efficiency from a rational point of view, people keep on screaming out "there is no time."

The word "efficiency" was generally used to describe the extent to which time, effort, or cost is well used for the intended task of purpose [12]. The competition that was based on efficiency, forming the management environment with the quote "the faster, the better," led to the conclusion that there was less time for the mind to rest. Now is the time of sensibility in which we should find the forgotten importance of the mind.

While the twentieth century was formed by materialism, production, and innovation, in the twenty-first century of sensibility and enjoyment, the heart becomes the key to open our feeling. Now, the interest of businesses does not only lie in innovation but rather in using sensibility to reach the hearts of customers. Although it can seem like a minor point, however, everything in the world is connected through relationships. People to people, people to objects, and people to nature, all are connected through relationships. The relationship between people is a relationship of minds, so delicate that the slightest mistake could make it shatter like glass.

Henceforth, the basic principle of economy is also not based on possession, but on "relationship." The founder of the Samsung Group, Byung-Chul Lee, claimed, "Do not become a salesman, but a business person who can reach the minds of the customers." The Korean word of customer means reconsideration, thus implying that the care of the

Fig. 2.4 "Customer First" written by the founder of the Samsung Group, Byung-Chul Lee (1910–1987), who was the pioneer claiming the customers' mind as the key of competition. By courtesy of ChosunMedia

customers' mind is important. No matter if an individual or a business, this needs sincere consideration to bring peace to body and mind (see Fig. 2.4).

The change of humanity can be divided into premodern, modern, and postmodern eras. The postmodern era is based on the feeling, popular culture, and nomadic items such as smartphones, cars, etc. If we are to divide the time into 10 years since 1980, the 1980s was the supplier-oriented era of insufficiency where the importance was put on quantity. While the 1990s was the customer-oriented era of abundance where the importance was put on quality, the 2000s is known as the time of choice and value.

While in the time of shortage, economic focus was on industries such as shipbuilding, oil refining, steel, etc., the keywords for the time of abundance were "light, thin, short, and small" characteristics of products. This is the time of choice, where keywords are "beautiful, enjoyable, soft, and creative" [41]. However, marketing campaigns of some Korean companies are still following gigantism based on size. For example, the largest department store Shinsegae Centum City in Busan still places Korea in the *Guinness Book of Records* for the world's biggest store.

The development of humanity is divisible into stages of hunting, farming, industrialization, information, and feeling. The lifestyle of humanity has been one of hunting for millions of years, followed by thousand years of agriculture. If the invention of the spinning wheel in England in 1764 was the beginning of the industrialization, it appears that this era has been followed by the information revolution. Today, information has become the infrastructure of any industry, which has led us directly into the time of feeling. This phenomenon is reconsidered in terms of the five elements: earth, wood, metal, fire, and water. Humanity has originally been a product from the Earth and has found ways to utilize the tree for lumber as well as for its fruit. Thus, in the Iron Age, more durable metals could be used for tools and weapons, after which people came to find how to produce energy. Now is the time of sensibility that is signified by the smoothness of water. The businesses in advanced countries today have already appealed for sensibility and have started to assign more and more key posts to MFA (Master of Fine Arts) graduates rather than MBA graduates in order to develop their products [42].

2.4 The Goods of Sensibility and Delicate Beauty of Slowness

History develops forming thesis, antithesis, and synthesis. A tribute band brings nostalgic dreams to life. For example, Dread Zeppelin plays Led Zeppelin songs in a reggae style with a lead singer dressed up as Elvis Presley, while Gabba performs the songs of ABBA in the style of the Ramones.

Most of today's Korean TV variety shows are casting too many idol stars to drive up ratings, but recently some singers, who used to play the acoustic guitar at the famous live café "C'est Si Bon" in the 1960s and 1970s in Seoul, have become popular again after they appeared in a TV show. The reason is that the audience can empathize with the singers performing with an acoustic guitar, which is able to make the original value last. The empathy of the audience to like good music sung by good performers is a value that never changes. However, at that time the live cafés disappeared without a trace, except for the café "Hakrim." Hakrim café started in Seoul in 1956 and has been a hideout for writers ever since. Although the owner of the café has to make money as well as to keep the atmosphere, he is still dreaming to continue the café, making it a place of harmony with 1,500 LPs, good coffee, and literary merit.

The "Cittaslow" manifesto suggests advocates keeping the traditional value of places, such as yards, cinemas, workshops, cafés, restaurants, etc., that are rich in soul [43]. Korea is very weak in keeping its traditions unlike Europe or Japan, despite its long history of 4,346 years, because the nation compressed the growth of centuries into a generation.

The power of products full of sensibility. The Danish home entertainment system maker Bang & Olufsen manufactures all of its products in the country without making any OEM deal. Targeting the 1 % of the upper-class market, they make high-quality items in small quantities based on orders received, so 60 % of the work is still done by hand. Thereby, their core competencies lie in the design field, and their home theater appliances are considered a top brand selling for around USD 121 thousand on the market for a home theater system set. In order to design and build only top quality products, they employ solely the top talents. Approximately one-third of the inhabitants of Struer where the Bang & Olufsen head office is located are employed at the company, also known as "the farm." The outside of the building is designed like a traditional Danish farm and outside there are sheep grassing on the field. The employees can enjoy nature and get inspired by their surrounding as well as bring peace to their hearts. Although it is the time of digital speed, they try to keep the sensibility of analog devices alive.

The global car manufacturer BMW has recently announced the expansion of female manpower and people of various ethnicities in order to secure diversity in the organization to make drastic changes within the company structure. The BMW group has been the number one car manufacturer by sales for several years. They have achieved new record highs for revenues and earnings in 2010, with revenues climbing by 19.3 % to Euro 60,477 million (2009, Euro 50,681 million), and Korea also posted double-digit growth: +74.7 % to 19,222. BMW claims its success to be based on three core elements: outstanding product, human resource management, and solidarity to feeling. Heike Schneeweis, the HR director at BMW who is the first woman taking such a high position, claimed women to be the future of BMW and guaranteed to lead the company with the intuitive power of a woman.

In 1598, during Japan's invasion in the last dynasty Joseon (1392–1897) in Korean history, people were taken to Japan in order to become potters. One of whom had settled there and his 14th descendant, Su-Gwan Shim, gave a lecture in the motherland, and he said that although they were taken to Japan, they built the Oksan Shrine and worshiped Dangun, the legendary founder of Gojoseon, the first Korean kingdom. However, it was unprecedented that in the invader's country, prisoners built a temple for their ancestor to maintain their identity. The potters have kept their identities after 400 years living in a foreign country, without changing their names into Japanese names. The family of Shim was able to keep their name in accordance with the will of their ancestors, "to find unmovable heart in the moving spinning wheel." Such is the national pride that the potters have kept for 400 years.

In May 2006, the author met with the 14th and 15th of Shim in the Shim Su-Gwan Kiln (see Fig. 2.5). The 15th Shim mentioned, "Traditionalism that has only passed down the skill does not live long. In

Fig. 2.5 The author dehyun sohn (*left*) and the 14th Su-Gwan Shim (*right*) who lives in Japan where his ancestor Tang-gil Shim was kidnapped in 1958. On the wall of his workroom, there is a frame written "With all my heart." By courtesy of the author, dehyun sohn

neo-traditionalism, the skill is passed down, in addition to making it likeable by the current audience, combining sensibility and the will to find a mutual understanding." He was proud that Japanese imperial families still purchase the products of the Shim's kiln and said what improves the dignity of his product to be the soul. He also added "I will live by keeping the love and affection of my ancestors' land Korea, by not forgetting to thank Japan; the country I was born and raised in."

Let's appoint ourselves for working slow. Being in a comfortable, quiet working environment, brainwaves come through much stronger and increasing creativity and reasoning. It is impossible to get good ideas in the busy and fast working environment. This teaching is called "slow thinking meditation." Some companies use the "siesta" (napping time in many cities in Spain) to let employees take a rest or provide yoga and other meditation facilities.

One day, the author saw a fantastic picture of the Russian scientist Natalia Avseenko swimming naked with the dolphins at the North Pole [44]. Avseenko stated, "Dolphins have the ability to sense the brainwaves of people. If a person is peaceful and the surrounding is friendly, his brain sends out Alpha waves. I was able to create these kinds of alpha waves through meditation." Furthermore, she said, "When I'm in the ocean and send out alpha waves, dolphins from far away sense it and approach me." When asking the former member of the Russia national swimming team why she swam naked with dolphins, Avseenko answered that she needed to be nude, since these whales don't like the touch of artificial fabrics, and recollected that they were smiling at her.

Bill Gates who has called for "creative capitalism" is a leading businessman who incorporated his meditation into management by employing a "Think Week," in which he spends one entire week alone to contemplate on the future of technology and then share these ideas with his employees. People need to go on vacation in order to fill their personal empty bowl. Every person needs around one to one and a half hours every day doing nothing in order to cut him off from his surrounding and spend time to think by himself. Thereby, meditation and breathing are the connecting links.

The breathing expert Hyo-Seok Suh, who runs an oriental medical clinic in Korea, suggested some secrets to pulmonary health: "take a deep and long breath," "use the nose, not the mouth to breathe," "laugh often to inhale fresh air," "maintain the air in the room fresh," "quit smoking," "clean the lungs through regular exercise and taking a bath," and "enjoy an outdoor spa in the forest or take green showers." He recommends breathing in deeply for 5 s and breathing out for 7 s in order to empty the lungs as much as possible. The reason for breathing slowly and deeply is to breathe out and completely empty the lungs to be able to take in more oxygen on the next breath. Breathing is the connecting link between the body and mind. The former Harvard-Yenching Professor of Chinese History and Philosophy and of Confucian Studies, Tu Weiming, claimed the core values for the peaceful coexistence of humanity lying in the constant moral training to clean the body and heart. The target of moral training is the benevolence, justice, courtesy, wisdom, and sincerity that are considered as the five Confucian core virtues.

2.5 Slow Life and Its Codes of Conduct

The general housing condition in cities is the "fast house" based on industrialized, standardized products, which destroy the overall environment [45]. In contrast to that, there is a new dwelling style, a movement to live properly called "slow housing." Out of these, an outstanding one is called the "straw bale house." Made out of straw blocks, it stands out in temperature maintenance and is able to provide a 40°C warm room in the middle of a 5°C winter. On the other hand, on a hot summer day of 40°C, it can maintain a comfortable indoor temperature of 24°C without having to darken the room using blinds. As the material is straw, it is possible to utilize the local products from the ground of each place.

The Harvard- and Yale-educated and American-born Western Zen monk, Hyon Gak Sunim [46], has agreed to the Cittaslow movement to match with the Buddhism lifestyle and architecture [47]. The architecture of Buddhist temples in Korea is known to be in harmony with its environment and is said to imply the philosophy of treating nature and humans equally. According to the Cultural Corps of Korean Buddhism, during the last 10 years, over 700 thousand people have experienced a temple stay in Korea, and among them 30–40 % of different religions enjoyed the charm of the Korean traditional culture through such temples. If this idea is applied to the traditional "old house," it can make visitors experience the "slow housing," which could be comparable with "Gite de France," a French hostel network.

Slow travel—the narrow and deep experience. Even in the tourism industry, "slow travel" is emerging as a new trend. Rather than flying many hours in economy class to see the picture of Mona Lisa behind bulletproof glass, people prefer to take the train to visit new local destinations and enjoy a rich experience. The characteristics of "slow travel" is one not requiring to travel very far compared to going overseas and not visiting various places at once. The preference of traveling by train rather than the airplane and by foot rather than by train is encouraged. Furthermore, while traveling one should be aware of the environment and visit slow cities or towns rather than busy and famous tourism destinations.

Today's tourists prefer to get storytelling items rather than souvenirs from their holiday. If the previous holidays provided a wide and shallow experience, "slow travel" delivers the "small and deep" experience. Examples of theme holidays are Bordeaux winery visits and prehistoric site explorations guided by a sommelier, artists, and marine biologists.

Another aspect of "slow travel" is the eco-friendly way of traveling. For example, if traveling from Paris to Marseille taking the high-speed train, it creates 10 kg of carbon emissions, while other methods, 187 kg by plane and 313 kg by car, leave marks of pollution in the environment. Therefore, "slow travel" is comparable with the farming of organic vegetables. More and more people are interested in "voluntainment" (volunteering and entertainment) programs which have the goal of helping other people on holidays while finding self-fulfillment and development. Recently, more people are interested in fair travel or responsible tourism.

Besides living and traveling, there are many things that we can do slowly. "Slow reading" is considered the spiritual "super slow." The scholar and great artist of Korea's later Joseon period, Jeong-Hui Kim (1786–1856), said, "The thousand books shall be in our heart, so they flow out to be paintings and writings." Thus, it is not necessary to read many books; however, those that are considered worth reading should be read slowly. The same goes for doing sports. Just like a snail, if one is able to lift weights within 20 s, it should be done slowly in 60 s. These "super slow" sports will strengthen the bones [48].

Fermented foods (such as soy paste, soy sauce, cheese, and yogurt), other products in season, and organic vegetables taste good and are healthy foods. It takes 15 min until the brain receives the signal that enough food has been taken in to the stomach, therefore, food should be eaten slowly. Anthelme Brillat-Savarin, the legendary gourmet in France, said, "The fate of a nation depends on the way they eat." In many countries in Europe, the average dinnertime lasts for 3–4 h and is considered "slow dinner," especially in Italy where slow foods start with "slow wine" that is slowly fermented to get the rich flavor.

It has been biologically proven that humans cannot live without love. Love is something that requires a lot of time and effort, and thus, it is also called "slow love." The Changdeok Palace located in Seoul, Korea, has a gate called "Donhwamun," standing for "mutual relationships are filled with love and respect." It remembers the "slow love" relationship between the king and his people, as well as the king and his government officers.

The Korean documentary film in 2012 about the late Catholic priest Tae-Seok Lee "Don't cry for me, Tonj" talks about a love story that moved the whole audience (see Fig. 2.6). The priest Lee who had graduated from a medical school in Korea travels to the wilderness in order to enlighten the aboriginal people by creating a brass band. It is not easy for him to become a doctor, since his mother is supporting a family of ten with merely a sewing job; however, he neglects all

Fig. 2.6 The late Catholic priest Tae-Seok Lee (*center*)—with his deep love and sensibility—became a star in millions of people's hearts in Sudan, Africa. By courtesy of ChosunMedia

honor and prosperity and chooses to go and live in Sudan, Africa, where two million people had died in the Sudanese Civil War. He follows God's words, "Whatever you did for one of these least brothers of mine, you did for me" and thus encourages patients who have been suffering from Hansen's disease in Sudan. He treats the people with his smooth and gentle actions; however, he does not know that his own life is ending due to cancer. He leaves this world at 48, telling the people "Don't cry for me, Tonj."

Walking helps you see a wonderful world. There is an anonymous poem saying, "The running man should walk/the walking man should stop/he stopped man should sit/notice the flower in the distance/look at the small flower closely/then you will see how beautiful it is." Walking is the first step of "slow life." People generally walk with a goal in mind; however, the Korean word for stroll refers to emptying one's mind without struggling for a goal. It implies to wander around in the side road or a devious route, walk along gazing around, and simply enjoy the nature while strolling around. In music there is a term "andante," which comes from the Spanish word "andar" (Eng., walk), indicating that the speed of slow refers to walking. To make a pilgrimage means to get off from the tourist bus or the train and walk the Earth with one's own two feet according to the compass of the soul.

After the invention of the engine, people have become lazier. Walking that used to be the least popular means of travel is becoming trendier due to the recent trend of "the slow movement" and environmental movements. Taking a walk revives the lost sensibility of the body that was taken from us by using the car and reminds us that we are "humans with two feet." Just by walking for 5 min, it is said that one can see a miracle.

However, many times people walk not for enjoying the benefits of walking, but for practicing sports, such as power walking [49]. Instead of enjoying walking slow, they only have eyes for the front, while sweating due to their fast walk. Although they have been involved in a busy and chaotic environment all day long, they throw themselves back into fast moving sports. When one spends some time on walking for exercise, a part of the time should be used to meditate and thoroughly check on one's body and mind condition.

Walking meditation is the easiest method to empty the mind. Going for a walking trip will make people notice the nature and people in their surrounding and explore the beauty of the world. The pilgrim road "Camino de Santiago (the Way of St. James)" in Spain is 800 km long and refers to praying while walking. In the poem "Walking Meditation," the Vietnamese Buddhist monk of high virtue, Thích Nhất Hạnh, writes:

Walking Meditation
by Thích Nhất Hạnh

Take my hand.
We will walk.
We will only walk.
We will enjoy our walk without thinking of arriving anywhere.
Walk peacefully.
Walk happily.
Our walk is a peace walk.
Our walk is a happiness walk.

Then we learn
that there is no peace walk;
that peace is the walk;
that there is no happiness walk;
that happiness is the walk.
We walk for ourselves.
We walk for everyone
always hand in hand.

Walk and touch peace every moment.
Walk and touch happiness every moment.
Each step brings a fresh breeze.
Each step makes a flower bloom under our feet.
Kiss the Earth with your feet.
Print on Earth your love and happiness.

Earth will be safe
when we feel in us enough safety.

As shown in Fig. 2.7, happiness comes on a bicycle. The car-related problems, such as congestion, noise, and air pollution through exhaust fumes, and obesity out of lack of sports can all be resolved by using the slowing-down bicycle. In the Netherlands, over 30 % of all journeys are made by bike, and all Dutch children bike to school, whereas 1 % of Americans uses bicycles.

Kyung-Hoon Lee, a professor of the School of Architecture of Kookmin University in Korea, claims in the book *Seoul Is Not a City* that in New York, Paris, and London, pedestrians come first; however, in Seoul the cars are given priority to the pedestrian. This was supported by the "Lonely Planet," a travel guide from England whose users in 2009 have voted the worst city to live in the world, and Seoul came in third. The world-renowned French aerial photographer, Yann Arthus-Bertrand, in the collection of aerial photography "Earth from Above" said, "The earth itself is one masterpiece. However,

Fig. 2.7 People who walk at 5 km/h (*left*). Bicycles that run at 15 km/h; an insect may be safe when it runs into a bike at that speed (*right*). These speeds are preferred at Cittaslow zone. By courtesy of ChosunMedia

Seoul was killed by the amount of cars in the city. You cannot see bicycles on the street."

Korea has started to establish many bicycle roads in a lot of areas in the city. It is a good thing that many bicycle ways are being made; however, due to cars utilizing bicycle roads and walkways, it is rather unsafe for the pedestrian to go for a stroll. In order to stop this, the downtown part of the city needs to be designed in a way where people come first as shown in Fig. 2.8. In most of the big cities in Korea, many cars are being utilized and take up a major part of the area, and so it is necessary to reduce the space used by cars. Also traffic should be reduced in certain areas of the city. In sacred places full of spirits, the approach by car should be limited and a "car-free day" established.

According to various studies in the world that investigated the influence the car had on the community, the more the car's speed was reduced, the more the relationship and contact within the community was encouraged [48]. The car was invented to save time; where did the saved time go? The car for dating, the karaoke equipment in a bar, and the cell phone to meet friends, all were inventions that were designed for pleasure purposes. However, the real pleasure comes from not depending on such technology, but enjoying life's invention itself by actively participating in life and pursuing enjoyment.

For walking or using the bicycle, one does not need to generate excessive energy and thus does not have to use it. This statement can be implied in the case of the recent nuclear crisis in Fukushima, Japan, where the production of electricity has been largely dependent on nuclear power plants. The Fukushima nuclear disaster was a release of radioactive materials at the Fukushima 1 Nuclear Power Plant in Japan. It is the largest nuclear disaster since the Chernobyl disaster of 1986. After the Fukushima nuclear disaster, most of the children spent the majority of their time indoors: home and school and school and home. They cannot even play with sand in the playground for fear of radiation exposure. Thus, a Korean organization held a fund-raising concert in Seoul, Korea. The purpose of the concert was to support programs allowing Fukushima's children to spend their holidays in Korea (see Fig. 2.9).

Fig. 2.8 "Bike hanger" (2 m × 5 m) designed by Korean designers, Jee-Yong An and Sang-Hwa Lee. By courtesy of MANIFESTO Architecture P.C

Fig. 2.9 "Korea-Japan Peace Concert" to support programs that Fukushima's children spend their holidays in Korea (*top*) and the children attending the program (*bottom*) (*Source*: www.ahimna.net)

At the first visit to Korea in 2011, Melissa Schilling, a new and renewable energy expert and professor in New York University, criticized the energy policy of the Korean government very strictly. Dr. Schilling said that in Korea there were more big cars than in Europe and criticized that those cars wasted too much oil. She claimed, "It is natural that the rise in oil prices generates the increase in using new and renewable energy such as wind power, however due to the limitation of the oil prices by the government it blocks the development of alternative energy." This implies that the Korean government should reexamine their nuclear energy supply policy and stop their energy pricing regulations.

The National Cittaslow Corea Network is currently running the Cittaslow club, "Jihe" (see Fig. 2.10), and its membership is open for anyone who is interested in being part of the slow life. We create a slow lifestyle that is better and happier and promote it. It leads to a more creative society and makes the world happier where nature, human beings, art, and economy live together. Therefore, Jihe enjoys the "carpe diem" lifestyle to its best. We develop business projects for

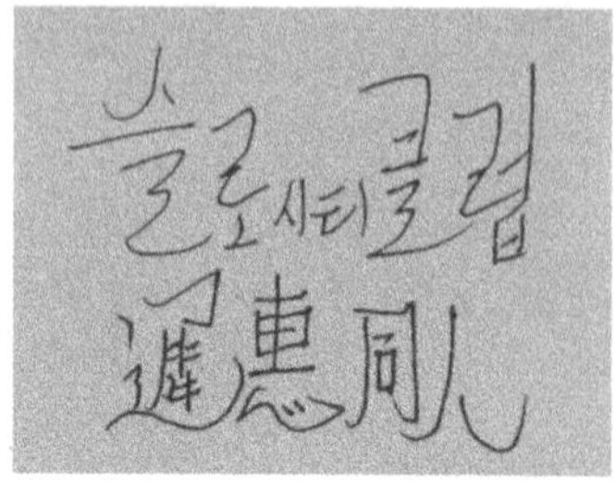

Fig. 2.10 The Korean word (*top*) and Eastern word (*bottom*) meaning the Cittaslow club, "Jihe" run by the National Cittaslow Corea Network. Written by author dehyun sohn

the public good and seek the common good through interaction between Cittaslow member towns to spread the Korean-style Cittaslow movement for the high quality of life and happiness of the public.

The members try to spread the philosophy of the Cittaslow movement and practice the 16 codes of conduct for a slow life:

16 Codes of Conduct of Slow Life from the NCCN

1. **Car-free Day per week**
   Walk, take the subway or ride a bicycle instead.
2. **Smiling is the best charming point.**
   Have a smile on your face because appearances matter and you have only one chance to make a first impression.
3. **Enjoy the food in a relaxed way.**
   Take a minimum of 30 minutes to have a meal, chew food 30 times, and eat 30 different kinds of food every day.
4. **Eat organic ECOfood that is in season.**
   Remove fast food from your table, and eat Ecological, Cultural, and Organic food.
5. **Do nothing every once in a while.**
   Take time to meditate and for silence.[slow meditation]
6. **A No TV day per week**
   Switch off the TV and read a book.[slow reading]
7. **Enjoy useful inconvenience.**
   Sometimes convenience is rather unhealthy.
8. **Wait for three seconds.**
   Don't press the 'close' button on the elevator, but wait for three seconds.

9. **Work *slowly, but surely.***
10. **Do not watch the clock on weekends.**
    Follow the clock of nature.
11. **Live your life like a traveler.**
    Take a 'slow travel' holiday, rather than a 'fast travel.'
12. **A 'No Shopping' day and a 'zero waste day' a week**
    Out of all animals, the human being is the only one creating waste, thereby destroying nature.
13. **An 'online sabbath' per week**
    Put your cellular phone away for offline analog.
14. **The 3Rs movement, a design for the twenty-first century**
    Reduce, reuse, and recycle
15. **Appreciate the traditions and wisdom from ancestors.**
    Restore our tradition to find a better future.
    Traditional wisdom is organic and sustainable.
16. **Live a simple life.**
    Make your life slow, small, and simple.

2.6 Slow Schooling and the Popularization of the Aging Society

Eight hours of work, eight hours of sleep, and eight hours of leisure time where we do whatever we desire; that is the perfect life. Children are considered "homo ludens," humans who play; however, recently there are many children who have lost the energy in life. Convenience is not necessarily a good thing. Happiness also comes from playing outside on a bumpy and uncomfortable ground. Young-Mo Ryu, a social thinker of Korea (1890–1981), stated, "The reason that God brought Jesus to the earth is to make him play like a child as being a child. Don't you think that the innocence of children, not knowing anything and purely enjoying life is the real living?" He always sang and danced since he believed that we had to behave like children before God. The "fast fast virus" of today does not only affect adults but children, respectively, giving them pressure to grow up even faster. Through the compulsion of having to do everything quickly, the childhood they deserve is taken away.

In Korea and Japan, children are forced to attend private cram schools in order to stay competitive. The books of such schools are so numerous that their schoolbag exceeds 15 kg, and thus, many children pull a trolley packed with books. The emeritus professor of education at Colorado University, USA, Maurice Holt, has started to spread the movement "slow schooling" in order to encourage educating children at a speed designed for children. Another alternative is the "homeschooling" program that liberates children from the continuous pressure of the fast education in the country and teaches them at home where they can learn in their own speed. In Canada 100,000 people follow this path of education, whereas in England 90,000 do so.

Everyone has to do what he likes. This is due to the distinct charisma of every individual. In some areas where a Cittaslow has evolved, they have arranged a "slow day," where children don't get any homework and can play as much as they want by getting out of the ordered life and simply enjoying their childhood. On that day, children can enjoy everything as slow as they want.

Now we need to focus on qualitative growth rather than on the quantitative growth of education. It can lead to the emergence of creative people such as the chairman of Microsoft Bill Gates, the chairman of Virgin Group Richard Branson, or the founder and CEO of Facebook Mark Zuckerberg. As a child, there is a method to learn how to think and feel certain things in life. However, ignoring this method and wanting to force the adults' method on children is considered a childish method [48]. Children should "ready, set, and relax."

A death of a senior means the disappearance of a library. If the number of people older than 65 exceeds 7 % of the population, we call it an aging society. The older population (persons 65 years or older) numbered 39.6 million in 2009 (the latest year for which data is available). They represented 12.9 % of the US population, about one in every eight Americans. By 2030, there will be about 72.1 million seniors, more than twice compared to the year 2000. People 65+ represented 12.4 % of the population in 2000 but are expected to reach 19 % by 2030 [50]. The aging of Japan is thought to outweigh all other nations, as the country is purported to have the highest proportion of elderly citizens; more than 25 % are over the age of 65 in 2013 [51]. The United Nations Population Division reported that in 2010 the percentage of the over-60 population accounted for

31 % in Japan; 27 % in Italy; 26 % in Germany; 25 % in Finland, Sweden, Bulgaria, and Greece; and 24 % in Portugal, Belgium, and Croatia. According to the United Nations, moreover, Korea's elderly population was 11 % in 2010 and is currently the most rapidly aging society in the world. Similar to a vampire's dilemma, people don't die as naturally as before and are beginning to face the time when the average length of life is 100. Now, Korea is also in the vampire's dilemma where people don't die. Therefore, a 100-year life plan is reconsidered with "30 years of education, 30 years of work, and 40 years of retirement."

The natural aging process is a beautiful gift to humanity. The death of a senior is the disappearance of a library. It is a "brilliant flower" that we have never been able to experience. In Cuba, despite the age of 80–90 years, such seniors have organized the Jazz group, the Buena Vista Social Club.

Any society which cannot embrace the phrase "seniors like seniors, children like children" is an abnormal and immature society. Seniors, just like children, are animals that cannot live on their own, and therefore, humans are designed to live in groups to survive. A mature society is one that accepts and tolerates the diversity of old and young people. The anthropologist Shinichi Takemura in Japan stated that the essence of humanity is the child and defined the old as slow time and young as affluent possibility [52]. There is the following consideration about age:

1 year—the age where everyone looks similar
12 years—the age where we learn the power of money
19 years—the age where we can watch any movie
24 years—the age where we pass down justice to the newcomers
48 years—the age where we statistically earn the most money
56 years—the age where we start disliking apartments
73 years—the age where we let gas in the presence of someone
82 years—the age where we have finished the human duties and just play

Someone asked the author if living slowly would imply a "restoration exercise"; however, this is not the idea. The "slow" movement is one that contradicts throwing away the past and only pursuing the future, but aims to put quality of life in the center though utilizing advanced technology and take the wisdom of seniors as well as the past history and experience to learn from and keep. To many people the mentoring of seniors is a help and guidance in life.

Many countries in the world like Korea and Japan are currently struggling with the problems of declining birthrate and aging population. In Germany, the more children people have, the more benefits they receive, whereas for single families, an income tax of 50 % is put on their earnings. Not long ago, the author has visited the electric installation equipment manufacturer Mirai Industry in Japan, where the CEO said proudly that employees receive a three-year maternity leave and employees of five kids receive 15 years of continuous income.

2.7 Points

We need to embrace the "slow life" in the speed of nature, which is the root of our thought, actions, and lifestyle. In order to live a new, happy life, we need a way to feel and think slowly. Considering the amount of fast-living people, there is rarely anyone who is living happily. People start to realize the sacrifices for living a fast life and the life on the Internet.

The city life is a technology-enhanced lifestyle, where the digital absorption is the enemy of depth. Although everyone has realized the danger of global warming, the amount of carbon emissions from the car is still steadily increasing. Korea ranked 8th in world energy consumption in 2010. "Slow life" is possibly the last opportunity for us.

People in current days always complain that there is "no time." The word "efficiency," which originated from using of machines, has been applied to humans stating "the faster, the better." Now the time of sensibility is the time to restore the importance of the lost mind. Sensibility is more important than reasoning and unconsciousness is more powerful than consciousness. Koreans are especially outstanding in "feeling" and have built up a clean and brilliant national character.

The twentieth century was all about material production and innovation, whereas in the twenty-first century the enjoyment of sensibility and the mind are the keys to the future. Businesses have to find the key that opens the customers' mind and establish a product that was developed with the charm of "slow" and feeling. Bill Gates has applied the motto "Think Week" and the charm of "slow" into the business.

The Slow movement is representing the activity of a slow life with NCCN's 16 codes of conduct. "Slow schooling" has started as a reaction of trying to substitute the learning method of children with the ones of adults and forcing them to overeducation. A society which cannot embrace the phrase "seniors like seniors, children like children" represents a difficult, abnormal, and immature society.

Slow Management and Small Businesses 3

Contents

3.1 CEOs that Live Without a Soul but Only with Their Body

Today, the CEO alone is not making all decisions in the company. The more gigantic, complex, and dynamic a company is, the more often it is that the CEO and a few executives sit together to make decisions, but it requires the involvement of the knowledge and ideas of all employees of the company.

In today's fast changing world, the CEOs use creative leadership to collect the imagination and creativity of the employees that creates a good business atmosphere where people enjoy working and show enthusiasm. Accordingly, CEOs should not be activists but act as party planners that organize and influence the energy in the surrounding with their persuasiveness and vision [53]. The example of the 1997 Asian financial crisis, where 16 out of 30 major

d. sohn et al., *Go Slow and Curvy*,
DOI 10.1007/978-3-319-14854-0_3

businesses in Korea considered to never falling into bankruptcy, shows the limitation of the vision and competence of Korean CEOs.

The first requirement for any successful politician or CEO is the degree of intelligence, a calm soul, and a peaceful mind [54]. You never know what might happen if you put too much responsibility on people who are nervous and busy at all times. The time where the body could survive without a soul is long over.

The world's richest man with assets of USD 46.6 billion, William Henry Bill Gates, the owner and founder of Microsoft Corporation, is known for the motto "Think Week" (see Fig. 3.1). Twice a year, he retreats alone into a villa in order to sort the absorbed ideas and strategic visions that he has taken into consideration from all of his employees in order to create a new strategy and ideas for the next 20 years to come. Bill Gates slowly read and opened the future in solitude of space and time through which today's company of Microsoft could be established. This is exactly the meaning of CEOs requiring time to be alone in slowness.

At the World Economic Forum in Davos, Switzerland, which took place in January 2008, Bill Gates was lecturing about his experience of "creative capitalism" [55] and was stating "large scale global enterprises need to include charitable work in their own company." Through managing their own beneficence, businesses should try to ease the unfairness or discrimination in the world and make a profit or guarantee more people a safer working environment and life stability through more job openings.

Actually, he is currently spreading noticeable charity work all over the Global Village. His "creative capitalism" is based on communalism such as fair trade, fair travel, and fair society and also on utilitarianism in accordance with economic justice.

Together with one mind. Let's take the example of global businesses. In Korea, Starbucks or Coffee Bean pays a 5–6 % royalty per cup to their US parent company, but sells a cup of coffee at the price 28 times higher than original price. In terms of Starbucks's tall latte index[1], one US latte at 2.45 dollars buys 0.5 French, 0.6 British,

[1] A representation of purchasing power parity published by The Economist that determines what a country's exchange rate would need to be in order for a Starbucks's tall latte to cost the same as it does in the United States.

Fig. 3.1 Bill Gates takes two, one-week "Think Weeks" a year to read and reflect on the world around him (*Source*: www.flickr.com)

0.7 Chinese, 0.8 Japanese, or 1.1 Peruvian lattes. In Europe, one cup of coffee costs around 1 Euro, and the farmers that produce the coffee beans are given around 0.03 Euros. When this reality was found out, the world brand Starbucks has been preferred by many people regarding Starbucks as conscientious corporation.

Starbucks markets with the slogan "Shared Planet," and they have a contract with 185,000 farmers across the globe who guarantee a good quality coffee bean in return for Starbucks paying above the world coffee price which helps the livelihood of such farmers. By the ethical purchase of fair trade certified coffee, Starbucks stimulates financial independence of farmers and claims to aim for a caring Earth and society. It is because of this kind of value management that more and more people have started to support Starbucks, planting the suggestion in their minds that it is good to purchase Starbucks coffee.

There is an idiom stating "beneficence is the ultimate happiness." Another similar phrase is saying "to those who give many good things will happen and they will find happiness." Some countries that had experienced rapid economic growth have begun to practice noblesse oblige. The Samsung Economic Research Institute in Korea did not only consider the "noblesse oblige" but also published the good infrastructure index in order to increase the corporate social responsibility [56].

Humans, compared to other animals, have the power of reasoning. Aristotle called the various goals of life as good. Furthermore, he added that out of those good, the most superior and ultimate form is "happiness."

What is a human being? Human beings are the combination of spirit, soul, and body. The philosopher Karl Marx (1818–1883) criticized the economy based on the philosophy of materialism, and Schumacher, who promoted the idea of human scale of decentralized, local economies, criticized the materialistic economy based on the spiritual virtue of life and particularly combined the view of economic spiritual values when he wrote his essay called: "Buddhist Economics" [13]. Among principles of Buddhist Economics are ideas such as minimizing suffering, simplifying desires, nonviolence, genuine care, and generosity; he discovered these core values in Myanmar where Buddhism has a close link to simplifying desires and nonviolence. Therefore, he claimed that economics without Buddhism is like sex without love. Although an economy without a spirit could provide humans with a temporary physical comfort, it lacks fulfillment. The essence of soul is "slow."

The essential front line of the "slow movement" is the enterprise. For today's people, work has become the major part of life and reduces time for other things significantly. According to the pioneer of the "slow movement" Carl Honoré, the Wall Street Journal hailed as "an in demand spokesman on slowness," while fast thinking is rational, analytical, and logical, slow thinking is intuitive and creative and yields rich and subtle insights. Time pressure leads to tunnel vision and prevents people from thinking creatively. In order to encourage creative thinking, we have to take the medicine called "slow," [12] where we are calm, unhurried, and free from stress and distractions.

Close the door of your office and close your eyes for 15 min. After that, bend back and take a deep breath. Even if working in a fast moving environment, doing this treatment will tranquilize the heart.

Sensibility wins over reasoning. The human mind is built on sensing and sensibility, and thus creative ideas are born through inspiration [57]. The Korean word "sinmyeong" has something in common with "ecstasy." Professor Gary Hamel at London Business School mentioned, "In a creativity-based economy, the most important thing is how enthusiastic the employees of a company are." Supporting his statement, he said that only about 10 % of employees are dedicated with their body and mind, the others are just present with their bodies. However, only if employees are taken care of and inspired, it guarantees a successful innovative business.

The energy of employees will become the strength of the enterprise [58]. If the basic desires of the employees are ignored and only focused on "more" and "faster," it will result in less and slower outcomes. Our body is poled that it requires a break after every 90 min. If a business wants to be successful, the physical condition of the employees and body energy should be maintained high, with deep sleep being the base of energy.

Furthermore, CEOs should keep acknowledging and praising employees in order to raise the sensibility energy of the organization.

The mental and spiritual energy are just as important. It is desired that people work every day from 8 AM to 8 PM without dropping their concentration. However, only if the CEO can deliver the high social value and company aims properly to the employees, he will get the best result.

The driving force of employees—dream and soul. The Japanese pub chain "Teppen" follows the philosophy that money can only move people to a certain degree. The driving force that inspires the employees is a dream and a soul. Facing every challenge, they kept asking themselves the questions "why we exist" and "what we are doing for the world." Only if the reasons for existence are clear, it creates a powerful partner that can break through any crisis; this is the soul.

Akio Yamada is the geek and chairman of the electric installation equipment manufacturer Mirai Industry Co., LTD, located in Kifu Ken in Japan and is known for his stinginess but also for his limitless efforts to serve his employees. He had been working for a while as theater director and has applied the principals of moving people to his employees. Only if the employees are happy, they can move the customer.

As shown in Fig. 3.2, his management philosophy is "we give to the customer first, and then earn the money," and he tells the

Fig. 3.2 Akio Yamada, CEO of the Japanese company Mirai Industry, tells the employees, "Get enough rest! Don't work for others! Do things you like!" By courtesy of the author, dehyun sohn

employees, "Get enough rest! Don't work for others! Do things you like!" Mirai is famous for its employee welfare policy: mandatory retirement at the age of 70, no layoff, overseas travel for all employees every 5 years, no night overtime, 3 years of maternity leave, 19 days off at the year-end holidays, 10 days off on golden week, 10 days of summer vacation, and the highest salary in the local area. Additionally, employees of this company each receive 140 days of holiday each year. During this time, he makes them slowly "think" for themselves. This results in employees raising the soul of the enterprise.

Although the business has produced more than 18,000 electric installation equipments, it has never duplicated one. He makes sure that his employees get enough salary and holidays in order to make them continuously use their brain. Rest first, and when you get tired of it start working. This is the "Mirai (it means future) management."

3.2 The Goal of Management Is the Happiness of Its Employees

Make your employees happy to make them dance. The American airline company Southwest Airlines is the only company in this industry which has not had any dismissals compared to 100,000 people losing jobs since September 11, 2011. Additionally, it ranked first on Fortune magazine's "100 Best Companies to Work For in America" in 1997. The cofounder and former CEO, who always puts people first, Herb Kelleher, said the following: "If you want to make your employees dance, then put their empowerment before money." As Southwest Airlines highly empowers their employees, they are given great latitude to keep customers happy (see Fig. 3.3).

The time when the CEO just needed to shout "Follow me!" in order to make the entire organization move as one is long over. Today, the leader is required to take care and build up the trust of the followers in order to receive their voluntary dedication to the company as response. Southwest Airline focuses on cheap tickets in accordance with "attention to detail." As a result, it could beat the big airline

Fig. 3.3 The legendary CEO of Southwest Airlines Herb Kelleher who puts his people first. By courtesy of ChosunMedia

companies that were just focusing on the "overall size." The management logic of "parts are bigger than the whole size" had an immense effect.

A manager is responsible for ensuring the happiness of people below him. The manager needs to prioritize the employee's happiness, since they dedicate major parts of their life to the company, which accounts for about two-thirds of a day.

The American software company SAS Institute Inc. has been in Fortune magazine's list of "100 Best Companies to Work For in America" 12 years in a row. The motto of this company is "slack" and is almost their competitive advantage [59, 60]. This company is famous for not having to work overtime, nightshifts, and without any dismissals and retirement issues. The working times are fixed to 35 h per week and guarantee a punctual end of working day. Furthermore, everyone including newcomers get their own office to work in. The CEO of SAS, Jim Goodnight, believes that contented cows give more milk and stated, "A company's most important asset is its people. Ninety-five percent of its assets drive out the gate every evening. Leaders should consider it their jobs to bring them back the next morning." If the employees, the creative capital of a company, can be used in an optimal condition, then the company will prosper (see Fig. 3.4).

"It's not true that productivity decreases by getting older," Jim Goodnight said, "I highly value the knowledge, experience and wisdom of older employees." SAS titled the company "the SAS institute" and the head office "the SAS campus." Jim Goodnight said that he tries to transform the image of the company from a general business to a continuously learning "school." SAS, after its foundation in 1976, has never had a loss in 33 years and in contrast had a constant average growth rate of formidable 8.8 %.

In addition, the founder of the multinational electronics and ceramics manufacturer Kyocera Corporation in Japan, Kazuo Inamori, who is also known as "the god of management," was criticizing the American performance-based management. "The management's goal is to pursue the happiness of all employees, both physically and mentally."

The freedom to live slowly. The less secure the workplace is, more people cling to work and moreover they do not go on vacation out of

Fig. 3.4 Jim Goodnight, CEO of SAS, states, "Contented cows give more milk;" the employees are the creative capital. By courtesy of ChosunMedia

anxiety to lose the job [61]. The criteria of job choice were also found to be money and job security.

However, we have become slaves of speed under the banner of industrial civilization in modern times and have been infected by the wicked fast virus. In 1968, the Club of Rome has published a report about the future of our planet, "The Limits to Growth," which has proven that the growth of the current route will not last long and given a suggestive warning. "After the Industrial Revolution, work has been controlling the life of people, which is insane, however, we desire to rule over our own lives and constantly seek freedom to live slowly." One of such miracles is the Cittaslow movement.

The Hyundai Motor Group in Korea has diminished the night overtime in 2012 and has announced to recruit 1,400 more people. The happiness of employees brings success for the company and ultimately creates a driving force for the national development.

Although there have been artists, poets, writers, and craftsmen who have followed the esthetics of "slow" during the industrialization, the route of business people was different. The reason that the word "business" comes from "busy" is not a coincidence but the attribute of the original meaning of the word. Additionally, the technological development has supported this phenomenon. This would be the reason that the phrase "if it's not busy, it's not a business" was established. Furthermore, in the Eastern character for "management," there are two characters of "fire" involved, which makes the meaning more intense.

Accordingly businesses have to be busy, making capitalism incompatible with "slow"? This is not true. Currently, employees prefer "slow" and support the tendency of a slow culture. Also, humans are originally poled to seek happiness, as well as to do things in order to help others, which at this moment requires slow business and slow management.

Let's think about good things for humanity, for society and the environment, for health, education, and culture. Contents in current time should be fun, meaningful, and slow for customers. Therefore, "good contents = fun + storytelling + slowness." It is necessary to create an afterglow effect that makes the customers happy, which does not have to be busy to sell well.

3.3 Sereno Leadership and Slow Management

Many experts say one should reduce the workload to work better [12]. In the end, to do less is to do more. If the working hours are reduced, it will lead to better work. In Norway and Finland, the working hours are much less than in other developed countries; however, their competitive edge is outstanding.

The words of the American author Helen Nearing (1904–1995) "Less is better than more" is a smart comment that support slow leadership. We have just recently lived through the hurtful experience of the financial crisis; however, this disaster was a result from the greed of humans, which has made the property and stock market as well as the money turnover of funds too fast in order to create more income. Let's follow the cultural trend of slow consumption. A mix of fast business and slow business or slow management should be in balance and harmony in order to create a new set of leadership that is an ancient future.

Three criteria to become a Sereno leader. In the book *The Stories of the Romans* by Nanami Shiono, she tells about the secret of the great Roman general Scipio Africanus (236–183 BC) who was able to defeat the greatest commander Hannibal (247–183/182 BC) in the Second Punic War due to his "Sereno leadership" [62]. "Sereno" is Italian and refers to serenity; thus a "Sereno leader" is a person who is slow, positive, and bright.

It requires three criteria in order to become a "Sereno leader." The first criterion is the scale of broad-mindedness, which means to accept the difference between oneself and another. However, the understanding and technique of Sereno come from a long endurance and experience.

The second criterion is detail, thus referring to the sensitivity toward others and careful consideration, that is, to have the time to listen to others' opinions before one's own and to be able to understand those opinions.

The third criterion is the smile. If one is able to maintain a good feeling, it reflects on other people as "shining bright." The American president Dwight David Eisenhower (1890–1969) was able to receive a high support rate through his "calming influence." In a survey about the political consciousness of the people by a Korean newspaper in December 2011, it showed a voting that politicians should be aware of with the title "the smooth leadership that can treat the wounds of the people." In this survey, it showed that 30.1 % preferred a strong leadership compared to 67.2 %, which is more than double that preferred to have a smooth leader [63].

There is another person who has adopted the Sereno leadership quite early. Khan Kublai (1215–1294), the fifth emperor of the Mongol nation, gave a secret order to the commander in charge who was about to go to war against the Song Dynasty of China and said "not to destroy the Song Dynasty, but to conquer it." He was able to make the imperial family surrender 12 years after the war had started without having to shed any blood.

Kublai had established an unparalleled empire including the entire area of China and middle Asia that was not comparable with any in the Islam or Europe. However, he did not follow the Grandfather Genghis Khan's strategy. Khan expanded his territory through speed battle and military power; however, Kublai completed the empire with his “slow management” style. Kublai employed the Sereno leadership “the slow, is the one who wins against the fast. Only if you know slow, you also know fast.”

The French luxury brand Hermes is in the top of the luxury category [64]. First, the price lies between ten thousand and hundreds of thousands of dollars. The reason for such a high market price is the artistic value and the effort of craftsmen for every single detail during the production. Hermes does not rely on division of labor, but one product is solely designed and produced by one person, and the whole process can be described as a single small boat on a tranquil sea. Every craftsman lives each day in solitude, while struggling to create art instead of a single product. Originally, the “saddle stitch” has been the traditional method of sewers for more than 150 years; however, using a cross sewing technique requiring both hands raises the product to a complete different level of quality.

The essence of luxury brands is that each craftsman has his own workshop and preserve to further develop unique production knowhow. The total number of craftsmen working for the leather section is 2,000 and shares the motto of “craftsmanship.” The fifth generation owner of Louis Vuitton Patrick mentioned, “A master craftsman is not a common technician, but an artist who is able to coordinate the head and hand to produce art.” Truefitt & Hill in London, UK, is the world’s oldest barbershop and has been operating for 207 years [65]. They are famous for offering the world’s best service and skill set and thus provide a service that has created three generations of returning customers.

Dr. Nam-Kyu Park from the Business Administration department at Seoul National University in Korea has observed the car production transportation pulley of the world’s number 1 car manufacturer Toyota. This pulley is the same that was used for the Karakuri puppets (using manual springs to set up the motion) during the Edo time (1603–1868), which runs without the input of any energy and is part of the transportation innovation. They originally changed the production and frame plan in order to use less energy; however, it did not only reduce the energy consumption but also decrease the noise level of the workshop. Taking the route of tradition and wisdom will always bring an answer.

Innovation without tradition will fail. Fissler is the world's best brand for kitchen appliances from Germany. The representative Markus Kepka was talking about the pressure cooker, "The pressure cooker was originally used in Germany to steam pork, and is the most integrated technology-intensive kitchen product from the whole catalogue." However, during the late 1960s and 1970s, the 3,400 nurses and miners that came over from Korea started using the cooker to prepare rice, with quite satisfying results, leading to many pressure cookers being sent to their families in Korea. This leads/led product innovation in response to customers in Korea and became the start of the Fissler's glocalization.

The philosophy of the product that has lasted through 166 years is based 40 % on tradition and 60 % on innovation, while maintaining the old frame. Fissler builds its products on the philosophy "Innovation without tradition will fail" and "Tradition without Innovation will disappear." In order to prepare for the next 20 years, and possible discontinuing the product line, they offer post-purchase services to exceed customer expectations.

Silicon Valley in America is considered as the center of entrepreneurship and the place where the energy of economic growth originated, thus representing a model, which is inimitable in the world. The most unique style of Silicon Valley is the network infrastructure rather than the hub structure of a company [12]. Since the business style is based on outsourcing, the organizational structure is placed very lightly. Unlike other big companies in the world such as US's DuPont, Japan's Mitsubishi Group, Hong Kong's Swire Group, Korea's Samsung, LG, Hyundai, SK, etc., that are present in many industries, small-sized ventures are represented based on ideas.

The second significant factor about Silicon Valley is the close watch on market changes and tipping points. This leads to many small changes adding together, finally causing one big change. Just like the quantum physics principle that states that the whole world is an impact of objects through moving from waves into particles, and before one knows the trend changes from one day to another. In that situation, one should prepare for recessions while the products are selling well. This is why people build up a vision for the next 5, 10, and 15 years based on their dreams.

The third factor requires not losing the amateur spirit in order to always stay creative and get new ideas. If a business falls into self-love of its own product, it will result in irrational bureaucracy. In order to develop new products that are differentiated and diverse, the business culture requires a fresh amateur spirit. This is why young people and a young-minded organization are important.

Fourth, Silicon Valley is free from the pressure of failure while large sums of incentives are granted when successful. Korea that was on the top of inventing mobile phones was surpassed by America through the smartphone becoming the world's leader. Also, the Korean social network service Cyworld had been the leader in the country and was soon surpassed by Facebook. Most of Korean smartphones have been manufactured by Korean big companies. Why was Korea expelled from the world's number 1 rank in the smartphone market by the USA and the nation's number 1 rank in SNS by Facebook?

Small businesses in Silicon Valley invest without the fear of failure. They consider that even if only 1–2 % out of 100 % of business investments is successful on a global market scale, it would make up for the failed 98 % more than enough. The reason for Apple surpassing Microsoft also lies in the investment of people getting on the Internet not over a browser, but through the use of applications and investing into smartphones which brought augmented reality into life. This was due to the culture of Silicon Valley.

Fifth, Dr. Sung-Jin Hwang from the Stanford Graduate School of Business has carefully watched the development of Silicon Valley for the last 20 years and has described it as 90 % culture and 10 % technology. A CEO who treats his employees equally and a corporate culture generous and open to immigrants are the source of different creative ideas. This is highly linked to the amateur spirit mentioned before. The free lifestyle and culture of sharing information and technology between universities, community, and ventures lead to the success of small businesses in Silicon Valley.

We live in a society where any big global company can be bankrupt tomorrow. It is necessary to take the right wave of global cultural trends and technology in order to gain sustainable competitive advantage from the lesson of Silicon Valley.

Fig. 3.5 Mark Zuckerberg, CEO of Facebook. The creative idea has provided the source of his fortune of USD 2.4 billion with the age of 27 (*Source*: www.flickr.com)

Many ventures turn into job opportunities, and entrepreneurs are regarded as heroes in the community. Bill Gates of Microsoft, Steve Jobs of Apple, and Mark Zuckerberg of Facebook (see Fig. 3.5), they all started their businesses in their 20s. Job seeking or creating jobs? The latter is the answer! Challenge yourself to create small and light businesses like in the Silicon Valley.

3.4 Slowness Is the Fundamental of Medicine and Education

In the past, the general trend was directed toward others-oriented consumption, how other people see one. This trend has shifted toward self-oriented consumption, in which the own person comes into the spotlight. In the past, shoppers were caught in the magical moment of purchasing a famous branded product, and shopping malls were attending to the customers with the attitude of "shop till you drop."

However, marketers today started to sell not only specific products but also a lifestyle. The New York Times has advertised places to see in Seoul, Korea, and has picked one that's called "10 Corso Como." If standard shopping was aimed at buying things that were needed, "10 Corso Como" is directed at fashion, art, and culture coexisting in a place to look around in a relaxed and enjoyable atmosphere to do "slow shopping." It generates a place where culture and art coexist and has brought up a different lifestyle, thus having no elevators to encourage people to slowly walk and shop with their own speed.

The identity of true education and "Slow Day." The role of a judge is listening to a defendant, who is the person directly involved in the crime, and the defendant's statement in court is critical to determining his or her destiny: guilty or innocent. Recently, "Slow Medicine" is coming in to the spotlight in the medical industry as part of slow management. Let's choose a place with more time for personal talk rather than having medical examinations and hi-tech equipment. A noted doctor is one who takes the time to explain things in detail rather than any other star doctor.

The treatment in a big hospital is equal to a conveyor belt. Doctors just look at the patient for some minutes and write a prescription. It seems as if they treat 40–50 patients this way in just 5–6 h.

The producer of "The White Jungle" Yoon-Hye Song handles the medical overcharge issue for the "30-second treatment." People in advanced countries, which have felt the disillusion of Western medicine, step back to complementary and alternative medicine (CAM) [12], which are coming into the focus in developed countries as part of "Slow Medicine."

In Korea, the traditional Korean medicine, alternative medicine, and folk remedies reach back to traditions; however, the philosophy of using complementary and alternative medicine embraces various medical philosophies in China, India, and Arab countries. As homeopaths there are methods such as herb treatment, perfume treatment, acupuncture, massage, yoga meditation, etc. In recent days, there are many hospitals that offer a package holiday combined travel with medical treatment.

The Virginia Tech Carilion School of Medicine is currently in the process of researching a fresh approach based on the statement "If you want to become a doctor, learn the art of speaking." To become a doctor, it is first required to learn how to talk properly to others. Additionally, it is necessary to be able to listen to the patient and his sensibilities in order to harmoniously communicate with him and his family and build an approachable character. Through this approach the Virginia Tech Carilion School of Medicine puts more emphasis on a medical student's personality rather than his academic performance. The applicants of medical school are required to have a total of nine eight-minute interviews which evaluate various different aspects. Eight medical schools in America as well as some in Canada participate in this kind of application process, whereas in Spain emphasis is put on a curriculum that encourages the doctor's bond-building ability with the patient.

Medical treatment is not regarded a dispensation but a service for the patient. Doctors tend to behave arrogantly like the all-knowing person, without taking the time to listen to their patients. Many people therefore complain that they haven't met a doctor who explains possible causes of illness or the treatment process in detail. In Korea, a person with outstanding medical skills is called "Eui-Sung," where the Eastern character "Sung" means holiness combined with the meaning "listen." Such patient-oriented medical approach was the base of traditional oriental medicine without having to rely on hi-tech equipment.

Education is foremost the learning process of finding self-identity (how to live) and a view of life before anything else [66]. Furthermore, this is a slow and progressive process.

The Latin meaning of "educare" is the "process to lead exerting one's ability." According to the Indian educator Sathya Sai (1926–2011), "educare" is related to human values. The word educare means to bring out that which is within. Human values are latent in every human being; one cannot acquire them from outside. They have to be elicited from within. Educare means to bring out human values. To "bring out" means "to translate them into action." Thus, education is the development of children's personality and teaching them how to exert their ability to live a happy life. However, currently most schools in Japan, China, and Korea do not aim for the happy child but for a child with the best grades.

The meaning behind the Korean word "Hak-Kyo" (school) is a product of two symbols. "Kyo" refers to the effort of treating a crooked tree to grow upright, through the exchange, counseling, and relationship between friends and teachers at a learning institute. However, since long ago, Korea had only focused on the symbol "Kyo" in "Kyo-Yuk" [eng. education], thus only concentrating on knowledge, making "Yuk" the counseling part very rare. It is a sincere process to raise a child. The origin of school is "scholē" in Greek, which means leisure. This explains the essential role of a school well.

According to the National Institute for Mental Health, suicide is the third leading cause of death among teens in the United States. In India, one-third of suicides are young people aging 15–29. In 2002, 154,000 suicides were recorded in India. In 2011 the World Health Organization reported that Korea has the highest suicide rate among the 30 OECD countries, having recently surpassed Japan's rate. In one day, there are in average 34 cases of suicide in Korea, out of which 50 % are people aged between 15 and 24 and relate to excessive education-related competition. The Korean word for university is "De-Hak," which refers to "great learning"; however, in order to enter a good university with high chances of a good workplace, the applicant requires a good school certificate, as well as TOEIC and TOEFL scores, overseas experience, internships, and other practical skills relevant for job seekers. It is the same in Japan and China. Hence, it's not wrong to claim that Korea's examination systems are a mass production of people with a very narrow outlook on life.

The US President Barack Obama admires the Korean education system, but it is considered as one of Korea's biggest social problems. Korea has a very high literacy rate of 99 % and according to "IQ and the Wealth of Nations" the average IQ of a Korean is 106, the second highest in the world. If Hong Kong is excluded, which is part of China, then Korea is virtually ranked top. A recent survey by Newsweek shows that Korean students have the longest studying times with an average of 7 h 50 min daily. In the Korean word for study "Gong-Bu," the Eastern character "gong" refers to the person who logically find the connection between heaven and Earth, a person who stands straight. Only if the fundamentals of "study" are all included, it is possible to apply them and create new things.

Face to face communication with an open mind. If we take a look at the academy award-winning film of 2010 "The King's Speech," it

Fig. 3.6 In the movie "The King's Speech," Lionel Logue (*right*) fixed the stuttering problem of King George VI (*left*) through open communication. (*Source*: movie.naver.com)

talks about the King George VI (1895–1952) who had a serious stuttering problem (see Fig. 3.6). During that time, Britain was preparing a war against Germany ruled by Adolf Hitler (1889–1945). In order to get across the unity of the Commonwealth, where 65 countries were involved to the people and military, the King's speech was critical. However, George VI did not have the courage to do so. In order to fix his verbal issue and become self-confident, he turned to the Australian speech therapist Lionel Logue (1880–1953), who continuously practiced and corrected him. As a result, he was able to give a formidable speech, and Britain finally won the war.

Lionel Logue did not study elocution at university, nor did he have a masters or doctors degree in education; however, he knew the importance of counseling in schooling. Coming back to Japan and Korea, although there are many universities and postgraduate courses available, counseling is still very rare. The problem is that there are no places where people can communicate face to face with an open heart and seek advice while listening to other experiences, because both professors and students are busy in their own business.

In some member towns of Cittaslow International in the world, they have decided on a “Slow Day,” where students don’t have any homework, but get the time to play and do whatever they want in order to feel like a child once more and enjoy life as a part of “slow schooling.” Now is the time, when instead of quantitative growth, qualitative development is more urgent in education.

3.5 The Small Businesses Form the Roots of the Real Economy

Small- and medium-sized enterprises (SMEs) are literally businesses that are smaller than big enterprises. In the USA, any firm from a small office/home office (SOHO) to a large corporation may be called an SME. More specifically, firms included in Russell indices such as Russell 2500 index and Russell Midcap index are classified as SMEs. In the European Union, a firm with (1) 50–250 employees, (2) annual turnover of Euro 7–40 million, (3) total assets less than Euro 27 million, and (4) not more than 25 % ownership by a large corporation may be classified as an SME. The International Chamber of Commerce (ICC) defines SMEs as having 100–2,000 employees. According to the Framework Act on Small and Medium Enterprises in Korea, a standard small- and medium-sized enterprise in the manufacturing business is defined as having around 300 people employed, with a capital of up to USD 7.4 million.

According to the Statistics Korea (2009), there are a total of 3,066,000 SMEs with 13,398,000 people employed compared to 2,916 big companies with 1,647,000 people. SMEs sum up to 99.9 % of total number of businesses in the country, with 87.8 % of people employed, generating added value of 49.5 %. This means that the majority of jobs come from SMEs. Therefore, such companies form the roots of our economy. According to the Samsung Economy Research Institute, the amount of exports by SMEs in 2011 added up to 32.3 % of total exports. However, if the exports of big companies are considered, where medium/small businesses were involved in production, the amount adds up to 54 %. There are approximately

20 SMEs per 1000 inhabitants in Germany, which is only about half of the EU average. More than 99 % of all businesses in Japan are small- or medium-sized enterprises; they also employ a majority of the working population and account for a large proportion of economic output. According to the Structural and Demographic Business Statistics of 2007, SMEs accounted for 99.8 % of total enterprises in France and 99.5 % in Germany and 60.5 % of total employment in France, 60.4 % in Germany, and 69.0 % in Japan and generated added value of 56.0 % in France, 53.6 % in Germany, and 53.0 % in Japan.

In order to build one car, it requires around 30,000 pieces. Starting with simple products, such as a bolt or a nut, up to core products such as engine transmissions, all are required in the car. Many medium/small businesses are involved in the production of such elements. The same goes for cell phones or home appliances. Thus, big companies rely on the economies of scale, being able to offer their products cheaper and on the quality of SMEs as their supplier. Many SMEs have a partnership relation with big companies as shown in Fig. 3.7.

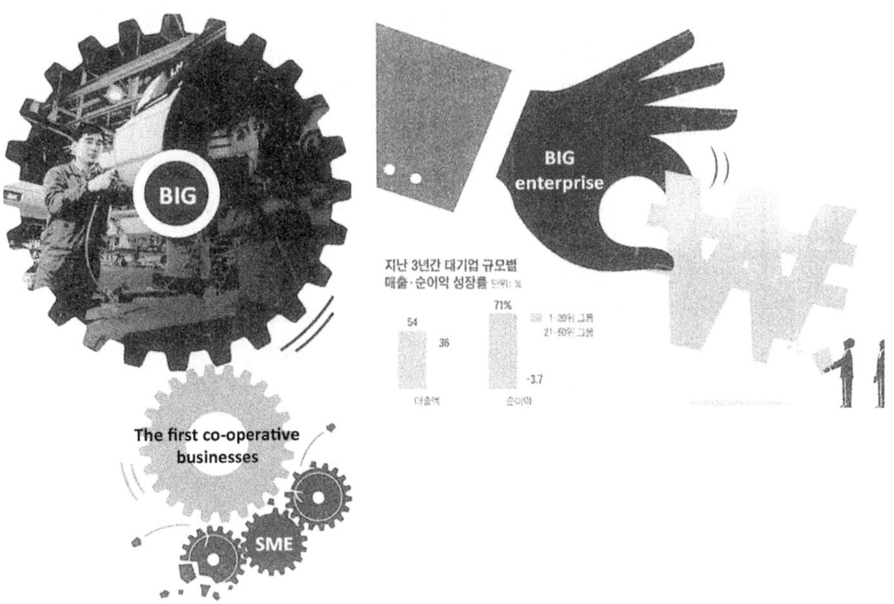

Fig. 3.7 Only if big enterprises cooperate with their strategic partners or other SMEs, the economy can survive. However, as the capitalism intensifies, the size and profit of big companies grow, whereas those of SMEs fall. By courtesy of journalist Hyeon-Guk Kim of Chosun Daily, Korea

The essential problem in the relationship between big and medium/small enterprises is the high level of dependency. This relationship does not include many big businesses, but only one, which is comparable with a caught animal in the zoo, which is fed and raised behind bars. Although the solution would be to get out of the food chain of big enterprises, cutting the food line would result in immediate death. Although such medium/small enterprises have had a high contribution and participation in the growth of big enterprises, the disadvantaged structure makes it hard to create employment possibilities for young manpower. If during the 1970s and 1980s the government hadn't supported an anti-market movement, the big companies today would've never grown to this extent.

The president of Hugh Max, the first venture in Korea, Dae-Kyu Byun stated, "We should be moving towards a 'young economy' where big businesses can fail, and medium/small businesses can become big. However, in Korea, big businesses become bigger and small businesses become smaller." In Korea, there is not a single medium/small enterprise that has become big by working with one of the chaebols that refer a Korean form of business conglomerate; thus it is recommended to find the market breakthrough overseas.

Small businesses and economic democracy. The reason why small- and medium-sized ventures need to be nurtured and supported lies in today's information age, where new technologies take the lead in the market, and thus it is required to have many ventures that are equipped with new ideas. Therefore, they should be supported in trying new methods and making adventurous decisions.

The Japanese research company, hidden champion[2] Macromill Inc., had 254 people employed in the company with sales of 7.4

---

[2] Hidden champions are relatively small but highly successful companies that are concealed behind a curtain of inconspicuousness, invisibility, and sometimes secrecy. A company must meet three criteria to be considered a hidden champion: (1) number one, two, or three in the global market or number one on the company's continent, determined by market share, (2) revenue below $4 billion, and (3) low level of public awareness (*Source*: http://en.wikipedia.org).

billion yen (USD 80 million) and a business profit of 2.2 billion yen (USD 24 million) in 2011, thus creating a profit margin of almost 30 %. Dr. Hermann Simon from Germany created "hidden champion," which is a small-sized enterprise, however, having a major competitive edge that it cannot be ignored on the market.

Furthermore, having a few hundred SMEs reduces the risk of the economy significantly. The president of the Switzerland Trade and Investment Promotion Daniel Küng noted, "It's unbelievable" that there are already over 300,000 Swiss brands made from SMEs, which creates a positive image for the country. Those businesses make up 99 % of total Swiss companies and are responsible for 75 % of the country's economy.

The smaller the company, the easier it is to put new ideas out on the market. According to the president Hans-Jörg Bullinger from the Fraunhofer research institute in Germany, the reason for Germany becoming number 1 in trading is foremost the result of the 1,350 small hidden champions. In Germany, 99.5 % of enterprises fall into that category. The total sales in heavy machinery add up to 43 %, with 70 % of jobs coming from this industry. Additionally, it was noted that 30 % of German engineers are college graduates. In the Netherlands, around 300 small businesses hold the bioproduct market, whereas even in Japan 50,000 small businesses are currently being supported by the government.

In the wave of globalization, where the wall of identity is about to fall, it is important that businesses like hidden champion remain alive. The Cittaslow movement makes sure that SMEs can get a competitive edge through regional advantages, differentiation, and high-class branding on the mass-producing market. Through such measures, a small city can become a livable city with depth and dignity.

The Economist, the global English-language weekly magazine, published an article titled in "Small is not beautiful" [67] and "Decline and small" [68] challenging Schumacher's "Small is beautiful" (see Fig. 3.8). Now, the eurozone crisis results from the sluggish productivity and loss of competiveness, which is related to a shortfall of big firms. Big firms can reap economies of scale, because they use far less cash and labor than small businesses, but they have their flaws, of course. They can be slow to respond to customers' needs, changing tastes, or disruptive technology. Eventually, size doesn't

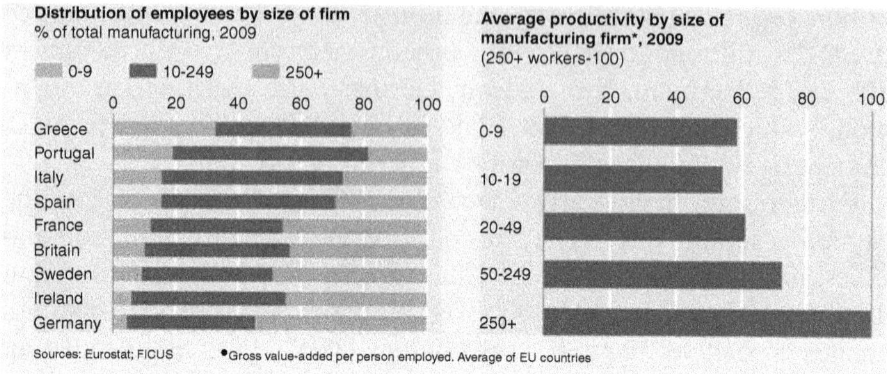

Fig. 3.8 General statistics of SMEs in the European region (*Source*: The Economist: Decline and small. March 3, 2013)

matter, but growth does. Although small firms created more jobs than big ones, most of them account for new start-ups, which create new jobs that did not previously exist. New small firms are more favorable for sustainable growth, because they can respond quickly to market demands.

There is a good example that shows the importance of SMEs. This is the "Tongkeun Chicken" (Tongkeun means "big bucket") dispute that took place in 2010. As the Lotte Mart in Korea announced to sell a whole chicken for USD 4.64 that is a third or a second compared to street price of fried chicken, it sparked the interest of many consumers.

The "Tongkeun Chicken" took the supplier's interest of selling many products as well as the consumer's desire of buying cheaply into consideration, creating happiness for both parties through this utilitarianism concept. When the author of "Justice: What's the Right Thing to Do?" Michael Sandel of Harvard University (see Fig. 3.9) visited Korea, he was interviewed about the "Tongkeun Chicken," whether the supplier, the consumer, and the small shops in the passage way were all taken into consideration.

Thus, he answered this issue, "If a big enterprise would reduce the price so much that all small businesses were to shut down, it would eventually lead to a monopoly and ultimately provide the power to put the prices up again. The immediate low price carries with it the risk of a high price tomorrow. Rather than having a few gigantic

Fig. 3.9 Prof. Michael Sandel, author of "Justice: What's the Right Thing to Do?" who said if small, independent producers become the focus of economy, it will create a greater democracy. By courtesy of ChosunMedia

businesses dominate the market, it will be more profitable to have many SMEs compete with each other, as it creates a strong democracy. This idea came from the supporters of the Associate Justice on the Supreme Court of the United States Louis D. Brandeis (1856–1941) during the 1920s and 1930s who took the lead in the Antimonopoly Act. Brandeis wrote that if SMEs were the majority, it would create a strong democracy. Furthermore, the curse of being oversized is a major concern."

"The core of a new German economic model is economic democracy," Christian Kellermann, the Director of the Nordic Office of the Friedrich Ebert Foundation, said in an interview with a newspaper during his visit to Korea. "Some employees in large companies earn extravagant wages, while many receive payment that is below minimum wage." It is the core of economic democracy that companies and unions should negotiate a wage increase. Although neoliberalists argue that economic growth cannot go with income distribution, economies will perform better with more even income distributions. The income equality contributes to increase spending of the public and demand, leading to economic growth. Such qualitative growth as well as quantitative growth will make the human society happy. Frontline employees are the people who understand the overall work on the spot and the needs of consumers. Therefore, their involvement in management can lead to the creation of innovative ideas, improving technology [69].

Such philosophic idea is also represented in the Framework Act on Small and Medium Enterprises in Korea that reflects that "the infiltration of big businesses in small trade markets is a violating process of trade" and should emphasize the public profit. If Korea is to open the USD 2 trillion trade market, there should not be any monopolies. It will be difficult to survive the intense trading war based on the existing paradigm which depends heavily on limited representative items of big businesses.

3.6 Creation Through Convergence

Smaller is more. The world-marketing guru Seth Godin once said, "In the rapidly changing 21st century, only the small survive" [18]. Prof. Rita McGrath from Columbia University Graduate School of Business criticized the Korean economy as "the largest Korean multinational conglomerate company Samsung has the lead in too many territories. In order to survive in the age of short uncertainty that can change the product and design instantly, it is necessary to bind hardware and software together, as Apple and Google did, to distinguish the product from others."

The great teacher of integrated leadership is the leading thinkers, writers, and commentators of the Korean Buddhist tradition Master Wonhyo (617–686) in Korea. His integrated philosophy is the "Hwajaeng (reconciliation of doctrinal controversy)" idea. This idea is the fusion of two totally contrasting generations. The logic is to integrate two to generate one. Thereby, the charm lies in the outstanding part of the consilience, creating differentiation and diversity.

The "Hwajaeng" teaching has a different logic compared to choosing between two sides and the thesis-antithesis-synthesis model. It combines two objects making a third new creation. It was found during the unity of the three kingdoms, Goguryeo, Baekje, and Silla, and showed the true forms of oppositions, between sacred and profane living, model, and customs during times when many disciples of the Buddhist priesthood were rebelling against it.

The Cittaslow International ultimately supports the integration of slow and fast, farming and industry, agriculture and city, and old and new things in a region and establishes order between them. Today, we desperately need the integration of small and big organizations, big enterprises, and SMEs. Only the small will survive and less is more. This is the resulting democracy. According to this principle, without the parallel growth of big as well as SMEs, the welfare of the people is just an empty statement.

The author of the book *The Power of Detail*, Wang Jungchu in China, described detail as having a strong power. Without detail, any management or individual is inadequate and put in the form "100-1 = 0." His theory of detail has originally been found in the Chinese tradition. "The difficult things in life all originate from easy things, respectively, all big things in life all originate from small things." This statement was the base in Laozi's Chinese classic text "Tao Te Ching (The Scripture of the Way and the Virtue)." The Eastern management, compared to the Western that focuses on things, looks for a solution to a problem on humanities. Even the Italian Renaissance sculptor Michelangelo (1475–1564) said, "Trifles make perfec tion, but perfection is no trifle."

The brand Louis Vuitton reconsidered the question of detail in their advertising campaign "Life is a Journey." (see Fig. 3.10). In this advertisement, the first American female astronaut Sally Ride, the second person who set a foot on the moon Buzz Aldrin, and the Apollo 13 Commander Jim Lovell came into the picture.

They are looking at the moon. If they would be looking at the moon, we should see their backs; however, we see their sides. Thus, they are not looking at the moon. They are looking at the sky. No, they have already conquered the moon and are looking for a new destination. They are looking at a different star. The next destination has been the premier value of Louis Vuitton for 150 years. The importance does not in the visual detail but in the meaning.

Less is more. The cutting edge commonly refers to a trend of the age. However, if looking closely at the Eastern characters, the

Fig. 3.10 Louis Vuitton's advertising campaign "Life is Journey." The models are not looking at the moon but a different star getting the message of "Life is Journey" across. The importance does not lie in the visual detail but in the detail in meaning. By courtesy of ChosunMedia

meaning of "the sharp end of an object" is written within. In short, if a small detail is maximized, it becomes the cutting edge. Therefore, detail is a powerful criterion. In the past, there was a poor couple, which had nothing to offer to the mendicant priest and looked up Guan Yin, the Goddess of Mercy in Buddhism. In that night, the couple had the same dream. In this dream, Guan Yin appeared and said, "Big things are small things of bigger things, and less things are more of lesser things." As the couple woke up, they did not understand the strange dream and went to see a monk to tell them about it, who wrote on a piece of paper "Oyujijok," satisfaction from having less, telling them to put this on the wall at home and keep on meditating and praying to Guan Yin with all their heart. It is said that after this incident, their wealth started to stack up mountain high, and even their children were very successful. This is also a teaching to be thankful and satisfied with small things in life.

"Oyujijok" is a Zen philosophical teaching, which refers to face reality and accept it, making your mind pure and satisfied with what you have and realize that "less is more."

This teaching is also valuable in current times. Miss Van der Roy who was born in Germany and lived in America was regarded the architect representing the twentieth century. She said, "The lesser ornaments there are, the richer the meaning. The more conservative the form is, the closer we get to the essence," putting more importance on essence instead of state, cause before result, and inner form before outward appearance. These conservative form and model language were her characteristics. Her philosophy was indeed "less is more." "Less is more," and "less is better."

3.7 Points

The higher up the ranking one goes, especially CEOs need to maintain a peaceful mind and a quiet soul. If we put the responsibility of a business, or a country in the hands of someone who is constantly nervous and chaotic, where would this world go? This is not the time anymore where the body can survive without a soul. There is a limit to move employees with only monetary value. The energy of employees is the power of the business, and the driving force that inspires them is the dream and spirit. The soul is based on "slow."

If a manager is responsible for the happiness of the people, then the goal of management should be to ensure happiness of employees. This will result in the success of the business and will ultimately serve as driving force for the development of the country. The aim of any business is to stay busy, and it moves with the argument "If we are not busy, we are not a business." However, the current consumers have the tendency to be slow consuming culture. Therefore, slow management will lead to slow business and slow leadership. Now is the time to develop a formula of "fun + storytelling + slowness."

If putting a strong, authoritative leadership in place, it is impossible to maintain high esteem. Let's remember the Sereno leadership: "the slow is the person who will defeat the fast. Only if you know 'slow,' you will know 'fast.'" The success factor of Silicon Valley lies in 10 % technology and 90 % culture. A culture that is slow, tolerant, and open will become the source of diverse and creative ideas. Additionally, new ideas will generate more job opportunities, develop a diverse industry, and attract more young

CEOs. Small businesses and cooperatives are small and beautiful because they are our neighbors like small veins.

Slow Medicine is similar to traditional Korean medicine. Emphasis in education needs to be put on values and identity. Without the slow concept, it will be difficult for educational institutes and schools to show qualitative growth, because schools are a fundamental of one nation's politics.

SMEs form the roots of economy. Although SMEs nurture a "young economy," it is not a jungle where one kills in order to survive, but mutual support is needed. Therefore, instead of creating an economy depending on a few big businesses, more small independent producers will establish a stronger democracy and succeed in the trading war. In this context, the idea of "Hwajaeng" by Wonhyo is supporting as an important principle.

4 The Slow Industry: Agriculture Revives Life

Contents

4.1 The Healthy and Beautiful, Everlasting Countryside

Agriculture is the representative of slow industry, traditional industry, and life industry. The reason for being the life industry is due to the general need of all living things for food. Now, life industry that is based on emotion is increasingly becoming a trend which cannot be avoided. The manufacturing industry prefers synthetic raw materials rather than natural raw materials and eliminates life in order to pursue perfect quality management. However, the fact that agriculture is the base whereas manufacturing is only a secondary industry has never changed, since people are able to survive without the manufacturing industry, while existing without agriculture is difficult. Therefore, it is necessary to find the proper harmony, where agriculture and manufacturing can coexist. The same goes for families where

d. sohn et al., *Go Slow and Curvy*,
DOI 10.1007/978-3-319-14854-0_4

housewives are in charge of the housework for family members, keeping them alive.

According to *Das Tempo-Virus* by Peter Borscheid, contrary to agriculture, commerce that depends on the manufacturing industry is a business that creates profit from speed. Traders order the production of faster ships at the shipyard in order to load, unload, and transfer their goods faster while building roads and bridges to accelerate the delivery [70].

Whether it is a civilized man or a barbarian, both are children of nature and not the rulers. Therefore, they should behave according to the law of nature. However, the civilized man has destroyed the majority of things on the land in which he has been living in for a long time.

Although the capital that provided by nature is enormous, nature was not recognized as capital or life and is being exhausted in a shocking speed. The pioneering organic apple grower Akinori Kimura in Japan, who has succeeded in growing pesticide-free "miracle apples," has become a star farmer throughout the world. The apples, which resulted in a surprising outcome of 14 years, are raised through a "natural cultivation" method that employs neither pesticides nor fertilizers. He trusted the power of nature, returning "miracle apples" to him.

He points out that the most important element in natural cultivation is the "soil," and good soil doesn't need fertilizer, so he leaves everything to the workings of nature. The earth needs to be vitalized first in order to vitalize the fruits. His agricultural practice is based on regarding humans as the fruit of the earth, and he teaches how to treat the land we are living in as the most important resource. Today more people weary of busy everyday life are finding escape in urban agriculture. According to the Solidarités International, urban farming is credited with producing 15–20 % of the world's food in 2011 [71], and over 700,000 Koreans in 2011 participated in urban farming, and there were 200 weekend urban community farms across the country as of 2010, according to the Agriculture Ministry. All the urban farming processes are based on manual labor, without machines, and creative farming can be done in collaboration with local farmers. Now, agritainment, a combination of agriculture and entertainment, enjoys great popularity for people who seek health and fun at the same time [72].

We can never win against nature. In the Amish Village in the State of Pennsylvania, USA, they ride horses instead of cars, light a candle instead of using electric light, and live without the Internet, television, or radio. Let's imagine this slow scenery. The dress reminds one of the nineteenth century. The Amish avoid using any mirrors whatsoever. Clothes are washed by hand rather than machines, and instead of using trucks, they utilize plows and horses in the traditional way of agriculture in order to seek the self-sufficient easy and simple life and live for a long time.

Just as the trickling water of a stream sounds beautiful due to the uneven stones it flows over, life also sounds wonderful when using bare hands and through the inconsistency of everyday life. Those people appreciate nature by being in harmony with it not by the illusion of possessing it. They do not compete with the neighbor but live alongside of each other. They adjust to Mother Nature's mercy and her speed. They know too well that they cannot win against nature.

Money destroys agriculture. Although "money" is not the element that vitalizes the Korean agriculture, the fact that culture is the number one power that enhances development is being ignored. In the past 17 years, the government paid out USD 107.5 billion to the Korean farmers. However, it was used for opening entertainment establishments such as Internet cafés and coffee shops [73]. If farmers who are given unearned money keep being dependent on it, it creates disorder. It is still being argued whether the agriculture is being wiped out due to the invested capital.

People and businesses are too focused on major cities, such as New York, Tokyo, Beijing, Seoul, London, etc. This has to be balanced with small cities in order to establish a suitable economic and cultural structure throughout the world. As mentioned above, we have to learn how to treat our land through the cultural structure, as this is deeply connected with our actions and lifestyle.

Schumacher introduced cultural practices into agriculture for a healthy agricultural development and mentioned that it requires a lot of effort in order to increase the standard and customs of agricultural practices [13]. In other words, without culture, agriculture cannot be improved and manufacturing cannot be settled; thus, the establishment of culture comes first. The statement "Agriculture forms the base of national existence" is a true core testimonial now and in the future. As such the consciousness of self-awareness and the stimulation of the spirit is the beginning of everything.

In South Carolina, USA, there is a saying "The best fertilizer is the footsteps of the owner," which means that farmers who walk across their land are looking and seeing what needs to be done and doing it; although many consider this refers to the heart of farmers, it is really their commitment to their land. Farming is the most faithful industry to principles.

The poverty of the rural community is more troublesome as poverty of thought rather than economic poverty. We have to change the direction of our thoughts. The small town in Korea called "Hae Nam" means "end of the land." We ask, "why do we call 'Hae Nam' the 'end of the land'?" But we see that it is due to its location, where the ocean ends and land begins. So it may also be called "Hae Buk" which means beginning of the land.

The lack of natural resources with material shortages is the second reason for the poverty of the rural community in Korea. The main cause of poverty is closely related to the problems of non-materialistic education and organization. Through proper education, the spirit can be stimulated and the culture can be changed.

"Exhaustion of the brain" based on sacrificing agriculture is devastating our lives. The creativity of humans and their imagination and thinking power are among the most important natural resources. Especially education is the one resource on top of the list and should be valued accordingly. Through the Cittaslow movement, we can get an idea and direction of how to live our lives. As education is a delicate process, it needs to evolve gradually and with care.

The inevitable truth is that any city will not be able to evolve properly if agriculture were to be destroyed. Agriculture has been severely damaged in the Philippines and Korea, even in the member nations of the OECD, which has lost its inherent sense of agriculture. That humans need to live with the natural cycle of agriculture is not a desire, but a necessity.

The Cittaslow movement is certainly different from other community reconstruction movements in the world, including China's Rural

Reconstruction Movement and Korea's New Village Movement. Where they are aiming for a village economy while destroying the community, the Cittaslow movement aims to empower average people to take the lead, while identifying the character of a region and searching for the quality instead of quantity of life by being a happy movement. By doing so, the goal of the Cittaslow movement is to establish a healthy and beautiful community.

Through this movement, the slow industry of agriculture has to be saved. Therefore, one way would be to follow, instead of going against the cycle of nature. Some countries in Asia, such as China, Japan, Vietnam, and Korea, have traditionally followed the 24 seasonal division points[1] in farming.

In the movie "Forrest Gump" (1994), there is the statement: "My mama always said you got to put the past behind you before you can move on." Tradition is the assembly of imagination, which is the door to new creation (Table 4.1).

4.2 ecoFOOD and Food Mileage

The people in today's big cities are suffering from an unparalleled degree of anonymity, social atomization, and mental isolation. Therefore, an alternative solution that was found recently is called "city agriculture." The amount of city farmers, who personally practice agriculture on weekend farms and vegetable gardens on the roof or on the veranda, is increasing. Such people don't just farm to enjoy the products but also enjoy the process of cultivating while helping to reduce the global temperature through their work.

City agriculture is possible within downtown buildings and unused land throughout the city as well as in shops. Furthermore, it can be done in secure production factories as well as in various education facilities where the vegetable gardens grow products for the school kitchen. The decision to join the increasing GIY (grow-it-yourself) community is therefore a process which changes a desolate city into a green city, thus changing the quality of life.

[1] There are 24 seasonal division points by which the solar year is divided under the traditional East Asian lunisolar calendars that matches a particular astronomical event or signifies some natural phenomenon.

Table 4.1 The 24 seasonal division points in 2014

Spring	*February 4*	**The Beginning of Spring**	*– Spring begins*
	February 19	**Rain Water**	*– It begins to rain*
	March 6	**Awakening of Insects**	*– Hibernating animals come to sense*
	March 21	**The Spring Equinox**	*– Day and night are equally long*
	April 5	**Pure Brightness**	*– It is warm and bright*
	April 20	**Grain Rain**	*– Rainfall is helpful to grain*
Summer	*May 5*	**The Beginning of Summer**	*– Summer begins*
	May 21	**Lesser Fullness of Grain**	*– Kernels plump*
	June 6	**Grain in Beard**	*– Wheat grows ripe*
	June 21	**The Summer Solstice**	*– It has the longest daytime and the shortest night of the year*
	July 7	**Lesser Heat**	*– Torridity comes*
	July 23	**Greater Heat**	*– The hottest moment of a year*
Fall	*August 7*	**The Beginning of Autumn**	*– Autumn begins*
	August 23	**The End of Heat**	*– Heat hides*
	September 8	**White Dew**	*– Dew curdles*
	September 23	**The Autumn Equinox**	*– The mid of autumn*
	October 8	**Cold Dew**	*– Dew is very cold*
	October 23	**Frost's Descent**	*– Frost descends*
Winter	*November 7*	**The Beginning of Winter**	*– Winter begins*
	November 22	**Lesser Snow**	*– It begins to snow*
	December 7	**Greater Snow**	*– It snows heavily*
	December 22	**The Winter Solstice**	*– The shortest daytime and the longest night of a year*
	January 5	**Lesser Cold**	*– It is rather cold*
	January 20	**Greater Cold**	*– The coldest moment of a year*

Currently, the urbanization in the world amounts to 50 %, and in 2011 the highest level is reported as 100 % in Singapore, Hong Kong, Macau, Bermuda, and Monaco, 93.5 % in Venezuela, 92.5 % in Argentina, 89.2 % in Chile, 84.6 % in Brazil, and 78.1 % in Mexico [74]. Korea also faces an excessive urbanization of over 90 % [75]. The acceleration of urbanization is a violation of nature, consequently causing a violation among people.

There is no other country where the health state is as bad as in rich countries. The Western society is struck with the 3F, food, fuel, and financial crisis. According to the World Bank, from 2008 to 2010 [76], Korea's energy use per capita was in the upper ranks among the OECD members along with Canada and the USA. Henceforth, home appliances are claimed to the 3E practice, emotion, ecology, and energy saving.

The current agriculture systems are too weak to cope with the negative changes from the front line of the industrial battle. Further, it is impossible to look at a bright future for agriculture based on modern farming methods. The fact that the profit from agricultural production is controlled by individual organizations or the government also needs improvement. In other words, it would be more desirable to have a closer link between the process of production and consumption. The supply from long distances is very destructive in terms of global warming and causes other side effects, such as uneven distribution, product costs, as well as an increase in carbon emissions.

Food mileage is the distance which the product travels from the producer to the consumer; the greater the mileage, the more damage the product causes to the planet. Grain, livestock products, marine products, etc., are representative of 9 imported items that were measured according to their area of production and distance from the consumer calculated as transport volume x distance. According to a recent study on the food miles of Japan, Britain, France, and Korea, Korea's food miles jumped 37 % from 5,172 km in 2001 to 7,085 km in 2010 [77]. Over the period, the three other countries all reported a decline; Japan from 5,807 to 5,484; Britain from 2,365 to 2,337; and France from 777 to 739. It can be said that Korea consumes more foreign agricultural products than any of the other three countries. However, the increase in food mileage does not only imply that the intake of fresh products is less, but the acceleration of global warming due to carbon emissions that are produced by transporting the products by ship and plane has a negative effect on the ecology of the planet.

The book *Meet the Mother in a Strange Garden* by the Korean garden designer Kyung-Ah Oh tells of the experience of a woman from the Lake District in England. One bottle of milk, one pack of eggs, and one bottle of apple juice were written on a piece of paper and put underneath a stone after folding it nicely. When she was woken the next morning by a rattling sound, she would find a bottle with milk, six eggs wrapped in paper, and a bottle of apple juice in front of her kitchen window. In England local food movement is currently spreading throughout various regions, thereby promoting the consumption of local products instead of importing less fresh items. This movement aims to decrease the cost of food through saving on transport costs and to revive agriculture by attracting farmers back to the locality who had left to work in other areas. The Lake District story beautifully shows us the promising future by "preservation" instead of "development," "fall back" instead of "head forward," and "slowness" instead of "speed."

There is a saying that there are three rivers in the human world. It refers to the rivers between the rich and the poor, the educated and less educated, and the city people and village people. These differences are partly due to the imbalance between the fast growing cities and shrinking agriculture. Therefore, the location of cities needs to be adjusted to only hours from farms in order to eliminate this balance.

Such points were considered when designating the Korean Cittaslow member towns such as Damyang, Namyang, Jeonju, and Sangju as well as the Italian Cittaslow Abbiategrasso. No matter how wealthy and plentiful, the health of the city (secondary producers) will always depend on the health of agriculture (primary producers). Accordingly, without the development of agro-industrial culture, the urban area and rural area cannot remain healthy. The following proverb is used in New Zealand: "Breathe in the cow dung, cockies—it'll cut your cancer risk."

Saving the important value of the autochthonous seeds. At the 2011 Davos Forum, one of the core agenda items was "FEW" (food-energy-water), in which it was formulated that food, energy, and water will become the most important keywords over the next

100 years. The Korean statement "Of eating comes having" underlines the value of enjoying delicious food in a slow manner.

We are born to eat food slowly. However, as life accelerates, people start to enjoy the convenience of fast food and even bring it home (see Fig. 4.1).

The ecoFood & cittaSLOW Forum was held in Cittaslow Jeonju under the theme of "A Beautiful Coexistence of ecoFOOD and cittaSLOW" on October 2014, and it aimed at introducing the term of ecoFOOD that stands for Ecological, Cultural, and Organic food, and encouraging the conceptualization and the use of ecoFOOD instead the words of local food or slow food. Participants at the meeting issued a statement, "ecoFOOD Manifesto" as follows: You are what you eat and humans can be said to be born to eat good food slowly, savouring every mouthful. Do you eat to live or live to eat? The latter is more appropriate because food is equal to life and life is an end, not means. The first purpose of the Cittaslow Movement is the improvement of quality of life in a community and member towns seek slow and deep ecology. Therefore Korea's member towns aim to promote ecological, cultural, and organic food (ecoFood) instead of local food or slow food. Humans unlike other living creatures don't live out their allotted span of life because of stress and bad food. It is expected that ecoFood is a new solution to save our food in a desperately dangerous situation. Our ancestors left excellent food culture like fermented food made from good ingredients that are obtained from an environment where Korea is surrounded by water on three sides with many plains and mountains and has distinct four seasons. Unfortunately, today Korean food has lost its authentic taste. As the most core element of cooking is ingredients, it is time to highlight food ingredients (Component), consumers (Guest), and producers (Owner), important players in food industry. Korea's member town members declare the ecoFOOD Manifesto at ecoFOOD & cittaSLOW Forum in Cittaslow Jeonju, promoting the use of the term "ecoFOOD" and enhancing its value as below.

1. Korean member towns keep in mind the inextricable connection between food and agriculture, and concentrate on heightening its value because agriculture is a representative slow industry and food is from environmentally sound agriculture.

Fig. 4.1 Italy is called "Eataly," where the concept of Slow Food was established against fast food outside Italy (*Source* www.flicker.com)

2. Korean member towns build a strong network through promoting ecological, cultural, and organic food and actively participate in the anabada (saving, sharing, exchanging, and reusing things) movement for a rational consumption.
3. Korean member towns use the term "ecoFOOD" and promote it through various initiatives and activities.
4. Korean member towns protect ingredients to save the authentic taste and flavor of local food.
5. Korean member towns focus on "eco", not "ego" to enjoy a happy life because food is vital for humans to live.

Nowadays, many aspects of the rural life are about to disappear. Rice paddies are pushed away by ginseng fields; the young generation leaves, leading to fewer students resulting in the closure of schools. However, just as the wind is always there without noticing it, the disappearing things also stay in a different place and do not vanish completely.

One of the most striking things that have disappeared is the native seed. Today, many countries in the world import fruit and vegetable seeds from foreign countries. According to the United Nations [78], the top major importing countries of fruit and vegetable seeds in 2011 were the USA, which imported USD 1,050 million worth of seed, followed by the Netherlands (559), France (430), Germany (347), Mexico (331), Italy (289), Spain (270), the UK (254), Korea (243), and the Russian Federation (225). Especially, Korea imports about 60 % of its demand for corn, wheat, and beans from major grain exporters.

In the twenty-first century, a time of food war and war of breeds, we should also focus on the development of nutritional breeds, new medicine, as well as fiber and bioindustry research to produce high-value items in order to secure our sovereignty and improve our export capacity through development. In order to prepare for the food breed war in the future, the Svalbard Global Seed Vault in Norway functions to preserve a wide variety of plant seeds in an underground cavern. Another example is the Millennium Seed Bank at Kew Garden in the UK. This is the largest ex situ plant conservation project in the world, and the focus of the project is on global plant life faced with the threat of extinction and plants of most use for the future.

They have successfully saved seeds from over 10 % of the world, and they aim to secure the safe storage of seed from 25 % of the world's bankable plants by 2020.

Bill Gates, who has participated in the World Economic Forum in Davos (2012), believes in the problem of the poverty of thought rather than the economic poverty and stated, "1 out of 7 people (1 billion) in the world fights with hunger, however, we cannot fight starvation without gene manipulation technology applied to grain to drastically increase production." Facing the opponents of gene transforming agriculture, he claimed that "we have to increase safety through experiments, if we cut royalties and the benefits for individual companies, we can bring forth many more benefits than disadvantages."

Dörte Schipper, the German author of the book *Den Tagen mehr Leben geben* (*Give the Days More Life*), asked, "If you were to die tomorrow, what would you choose as your last meal?" The majority of people were choosing "their mother's or grandmother's food at the end of their life." The Swedish Cittaslow Falköping focuses on locally produced, ecologically sound and happy animals, whereas Koreans value their domestic, native food, because food controls their health and national characteristics.

Agriculture can nurture a country. Many countries in the world have been defenselessly invaded by multinational GMO companies. Soy, corn, wheat, potatoes, beans, etc., have been dominated by the GMO. The self-sufficiency in wheat in many countries has been low, for example, in Korea which produces 1.7 % of its own wheat. According to the US Department of Agriculture data, the world's top ten wheat importers in 2008 were Egypt (9,500,000 tons), Iran (8,500), Brazil (6,500), Algeria (5,600), Japan (5,500), Indonesia (5,300), Morocco (4,000), Iraq (3,700), Nigeria (3,500), and Turkey (3,500). Most Asians enjoy various kinds of noodles. The reason behind this is the slurping sound it makes when sucking in the noodles, which stimulates the happy hormones and increases the taste through the auditory sense. Furthermore, the reason for serving noodles in banquets also originated from the length of noodles which indicate best wishes for a long and happy life. The farmers in the Cittaslow areas need to revive the sovereignty of the native and local wheat breed.

The breed wars, climate change, impact of FTA on agriculture, etc., all are forcing heavy changes on the agricultural industry. Now is the time to choose selected products from grain, fruit growing, vegetable, and livestock industry and put policies into place to reduce potential dangers and make them available for agricultural export by re-regionalizing the products and to establish its identity and put emphasis on the distinguishing feature of items [79]. In the Negev desert in Israel in the Arab area, an irrigation facility was established for breed improvement and is said to have the potential to nurture the whole of Israel [80]. These organic products, such as chili, eggplant, and tomato, form 95 % of agricultural exports to Europe.

4.3 Cittaslow^{Slowcity}: A Slow and Deep Ecology

Every living creature has its own right to live on Earth. The cow is not meant to live to provide milk for humans. We cannot simply measure the value of a cow on this role alone. Every natural element has its own individual value. Even things that do not seem to play any particular role for humans, such as weeds for grass, scrub for trees, and miscellaneous grains. If humans decide of its extinction according to its usefulness for human beings, it is an arrogant attitude.

Even more, humans gain access to control over life by engineering the genes of species and shortening the natural process of evolution. This is nothing but human tyranny. The Cittaslow movement pursues a slow and deep ecology. Just as mentioned previously, the saving of the native breeds and the consumption of native food are measures to protect the natural environment.

According to Hannah Nordhaus, author of the book *The Beekeeper's Lament*, the honeybee asks, “Human, are you aware of the sin that you are putting upon nature.” The “honeybee” is vanishing. The naturalized Korean citizen of German birth Dirk Fuendling, whose Korean name is Bin Do-Rim and who lives in Korea's Cittaslow Damyang, blames all sorts of pesticides, electro-magnetic waves of wireless phones, viruses, and bacteria for the collective death of the honeybee, all of those causes being part of environmental problems such as an abnormal temperature. Albert Einstein (1879–1955) predicted, “If the bee disappears from the surface of the earth, man would have no more than 4 years to live.”

Out of the plants on the planet, 1/3 are entomophilous plants, which are fertilized by insects, and the honeybee represents 80 % of such. In conclusion, the decline of the honeybee is a significant factor that threatens the basic supply and demand circle and thus our survival.

Humans take away the required time and space that is needed for life in order to create fast food. In chicken and pig farms, they inject antibiotics in order to make the animals grow faster. Just 200 years ago, it required 5 years for an average adult pig to reach 130 lb. Today, it barely takes 6 months, sometimes even only 3 months [81]. Similarly, in North America salmon have been genetically modified so that they grow 4–5 times faster than the average unmodified salmon. The small farms are pushed away by big commercialized farms, which produce cheap, fast, and standardized products which many consider tasteless and poor in nutrients.

The Slow Food Manifesto that was published in November 1989 in Paris, France, states,

> Our century, which began and has developed under the insignia of industrial civilization, first invented the machine and then took it as its life model. We are enslaved by speed and have all succumbed to the same insidious virus: Fast Life, which disrupts our habits, pervades the privacy of our homes and forces us to eat Fast Foods. Our defense should begin at the table with Slow Food. We should rediscover the flavours and savour the taste of regional cooking and banish the degrading effects of Fast Food.

Slow Food is considered loving small things, living without hurry, and eating food that grows locally, so it seems like an old enemy to the global capitalism.

Slow Food respects the speed of nature. The three criteria of Slow Food are good food, clean food, and fair food. Whereby, good food refers to the non-standardized, non-artificial, and non-preseasoned food aiming for the original taste. The W. K. Kellogg Foundation in America developed the indicators for healthy, green, fair, affordable food and food system and seeks "Good Food for All."

Eat less meat. Korea and Japan, countries which rank first and second in terms of food mileage, rely extensively on foodstuff from overseas. However, they do have some renowned local foods. For example, the reason for Japanese Shorthorn cattle are prized by Slow Food aficionados is due to the feeding regime. The cattle spend the summers grazing on plentiful grass in vast mountaintop pastures instead of eating imported feed grains in stalls. Thus, they are produced locally and fed on local grass from the Iwaizumi forest, which results in low food miles.

The exceptional character of the beef produced by these cattle is the low fat and high protein content. The taste of beef varies depending on what the cow was fed and has a different taste, texture, and price depending on the part of meat. The Korean native cattle that are corn fed have a good marbling, whereas the grass-fed cow from Australia or New Zealand is characterized by a good chewing texture.

The problem is that the majority of corn is GMO. On the contrary, in Cittaslow member towns in Europe, the natural character of meat is being maintained by reviving the vanished varieties of animals and plants and concentrating on the production of bio-pork and bio-wines. In Mendrisio, the Cittaslow of Switzerland, the cow is sent through the Alps for 3 months to assist the animal's welfare and improve the cheese and milk produced and to ensure that the meat is in a better condition.

The outbreak of the foot-and-mouth disease in the UK in 2001 caused a crisis in British agriculture and tourism. In Korea in 2010–2011, a serious outbreak of the foot-and-mouth disease led to a mass cull of approximately 12 % of the entire domestic pig population and 107,000 of three million cattle. Just a few months before, Britain had experienced the mad cow disease, which was caused by the impatience and greed of Europe to raise its livestock by feeding the grass-eating cow concentrated processed animal feed.

According to Paul Roberts, the author of *The End of Food: The Coming Crisis in the World Food Industry*, the foot-and-mouth disease and the avian influenza were both caused by the desire to purchase cheaper, more convenient products and to consume livestock products in large quantities, resulting in a crisis in the cultivation and food distribution process. In order to grow livestock faster, they were put in sheds instead of being in their natural environment, which created the best conditions for major diseases to grow exponentially. Furthermore, antibiotics were utilized, which raised the resistance of bacteria and increased their attack rate.

It is recommended that the wisdom of our ancestors is followed in order to prevent outbreaks of the foot-and-mouth disease. Sheep, cattle, and swine that have symptoms, such as higher temperature or loss of appetite, cause the foot-and-mouth disease (also called hoof-and-mouth disease). In Korea, the traditional procedure during winter times is the provision of straw on the floor of the barn in order to maintain a warm temperature. If this method is not sufficient, cows are provided with straw mats in order to keep them warm. More importantly, the straw that was cut with straw cutters was used to make fodder, and bean pod, corn, and rice chaff were used to make a hot porridge that the cattle consumed. It is in the flow of nature to feed the grass-eating cow its natural nutrition.

Mold, the first Cittaslow in Wales, UK, recommends eating less meat and instead consuming more vegetables, pulses, nuts, and grains to save money, improve health, and reduce carbon emissions [82]. In the Bible, Genesis Chap. 1, verses 29–31, it says,

> And God said, Behold, I have given you every herb bearing seed, which *is* upon the face of all the earth, and every tree, in which *is* the fruit of a tree yielding seed; to you it shall be for meat. And to every beast of the earth, and to every fowl of the air, and to every thing that creepeth upon the earth, wherein there is life, I have given every green herb for meat: and it was so. And God saw everything that he had made, and, behold, *it was* very good. And the evening and the morning were the sixth day.

According to the order of creation, a harmonious peace relationship is undoubtedly found in a vegetarian lifestyle.

The food self-sufficiency rate in the world is decreasing, especially among crop-producing countries such as Japan and Korea. The figure is the lowest among the major industrialized countries. If Korea was to exclude its historical livestock records, the degree of food self-sufficiency would add up to just 21 %, which shows its high dependence on imported products and thus an extreme creation of carbon emissions. Recently, the vegetables raised in vertical farms[2] in cities introduced an alternative direction. It represents one way of reducing carbon emissions by the direct distribution of local agricultural products. Therefore, such areas are continuously trying to develop

[2] Vertical farming is cultivating plant or animal life within skyscrapers or on vertically inclined surfaces.

solar power, wind power, and bioenergy to create new energy sources and stop the production of greenhouse gases.

4.4 The Beloved ecoFOOD, "Soy"

Former professor Hak-Su Ahn from the College of Agriculture at Korea University claimed that the world's best ingredient is the bean. Although we do not know the true full history of the bean, just by looking at the Eastern character "Nong (agriculture)," we can see that it is composed of field and bean and thus infer that the bean has been eaten for a longer time than any other grain in Asian countries.

The Korean native bean (see Fig. 4.2) contains 48 % protein, 25 % fat, and 25 % carbohydrates and is thus a top quality piece of protein [83]. Especially urokinase, a serine protease (an enzyme, present in the blood and urine of mammals) found in fermented soy food, removes blood clots. The Korean Institute of Technology Research (KAIST), a public research university, has even developed a livestock wastewater treatment solution using fermented bean. They succeeded in using the bacillus from the traditional Korean sauce "Cheonggukjang (fast-fermented bean paste)" in livestock excretions to create clean filtered water in just 2 weeks.

In China, Japan, and Korea, it was not common to consume protein via animal products, and thus, people consumed vegetable protein via beans [84]. The bean does not take account of any status or class and is thus a product that was traditionally enjoyed by everyone for a long time. Even during winter times, when vegetables could not be grown, products, such as soy paste, soy sauce, chili sauce, and "Cheonggukjang," could be made using germinated beans, and non-fermented products, such as bean sprouts or tofu, were enjoyed likewise. Nonetheless, today, soy sauce is produced in large amounts without regard to achieving the perfect balance of beans, salt, and microorganisms due to the impatient character of people, and thus, the true essence of "seeded soy sauce" has been lost.

Happy beans: soy is better the longer it sits. Korea's agricultural book *Enlargement of Forest Economy* (1766) stated that each family should spend more time in making better soy sauce, because it has an impact on meals for 1 year. In the past, farmers used to save three

Fig. 4.2 It takes 3 years from boiled beans and fermented bean lumps to Korean sauce, "Jang." The fermented sauce business, Gosirang in Boeun-gun, Chungcheongbuk-do, Korea (By Hee-Ki Min, Courtesy of Homeliving & Style)

seeds in every hole when cultivating soy in the order that birds eat one, insects eat another, and they grow the other.

The bean has many siblings. Just in Korea there are 100 different kinds of bean; however, although being grown in the country of origin, it is assumed that almost all bean products contain GMO now due to the commercial use of these GMO beans in the production of soy paste and soy sauce. The culture of herbicide-resistant GMO beans by multinational companies does not let any weeds sprout and does not attract any insects or birds. There is a verse in a popular song saying that the woman who was weeding a bean field doesn't need to weed any grass today.

In Tokyo, the number of high-rise buildings and fast-food chains is overwhelming, while the food self-sufficiency has dropped below 1 %. The towns that are preserving the self-sufficiency percentage are not big cities, such as Tokyo, but the small, unknown villages outside of the urban area. In such areas, people try to sustain the nature and the importance of local products, while trying to regain dignity.

Soybean paste, soy sauce, tofu, and natto (Japanese fast-fermented bean paste) are all based on beans; however, the self-sufficiency degree is below 5 %, which signifies that the majority depend on imports, making the native Japanese bean even more valuable. In the Slow Food village, Onogawa, in Japan, 30 farms involved in bean sprout farming have cultivated the black bean for over 150 years. The significance of the autochthonous black bean is the flavor and strong taste compared to imported beans.

In Korea's Joseon Dynasty (1392–1897), soy sauce used to be ripened for 10 months with high sincerity and was usually placed in the middle of the dining table. An owner who has operated the traditional Korean restaurant in the center of Seoul for 23 years mentioned, "The base of Korean cuisine is its side dishes. In order to make the side dishes tasty, using the right ingredients such as soy sauce, soy paste, chili paste, water and salt are of utmost importance." The main material for making soy sauce, soy paste, and chili paste is the soy, which was the reason Koreans were once called the soy people.

The taste of Korean dishes is the soy taste; depending on the soy, the flavor changes (see Fig. 4.3). As the soy is a fermented product, it is close to a perfect protein item. A presentation from the medical

Fig. 4.3 Traditional Korean sauces made from soy add deep flavor to the Korean cuisine (*Source* www.pixabay.com)

research team, Chonbuk National University in Korea, had thoroughly investigated the effect of soy based on a clinical assessment and found that the key to everyday problems, such as obesity and diabetes, lay in soy. By reducing the fat in the body and increasing muscle volume, soy is the best product to produce a fit body.

Bibimbap[3] (see Fig. 4.4) from the Korean Cittaslow Jeonju is well known as one of the three best dishes together with the cold noodles from Pyongyang in North Korea and Tangban (boiled rice served in soup) eaten during the Joseon Dynasty. Above all the essence of Bibimbap lies in the thorough diversity, which gives a rich taste that is shared within the community. Just as the German philosopher of romanticism Johann Gottfried Herder (1744–1803) called the coexisting communities "The various flowers in God's garden,"[4] the various vegetables that are mixed in Bibimbap (mixed rice) give the dish a rich and whole new flavor.

The essence of Bibimbap however depends on the quality of beans that are used for the chili sauce. Recently, the popularity of Bibimbap has spread globally, finding many people who enjoy the dish as a healthy diet, even creating a Bibimbap burger in the middle of the best gourmet dishes in New York, USA. The Bibimbap burger has even been selected the best hamburger in America's best burger contest of 2011.

In the Chinese literature "The History of the Three States before its collapse," it states, "Koreans assemble to drink, sing and dance for a whole festival." Briefly, it may be said that people enjoyed the drinking culture.

Especially Koreans enjoy drinking, singing, and dancing compared to other nations in the world, which has always been part of the Korean culture. If we take a look at the Korean, Chinese, and Japanese alcohol culture, in China, where the cold north wind of winter blew fiercely, people consumed pork with the over 40 % proof

[3] Bibimbap is a signature Korean dish. The word literally means "mixed rice." Bibimbap is served as a bowl of warm white rice topped with namul (sautéed and seasoned vegetables) and gochujang (chili pepper paste).

[4] If every one of these nations had remained in its place, the earth might have been considered as a garden, where in one spot one human national plant, in another, others bloomed in its proper figure and nature.

Fig. 4.4 Mayors of international Cittaslow member towns making bibimbab at "eco-Marine Town Forum" in Cittaslow Wando, Korea on April 2014

Chinese Kaoliang. In Japan, on the other hand, in a hot and humid climate, people drank "Sake (a traditional Japanese alcoholic beverage made from fermented rice)," which although lower in alcohol had a faster intoxication effect. In Korea, people have enjoyed "Makgeolli (a traditional Korean alcoholic beverage made from fermented rice)"[5] in a bowl full of compassion with some chili and cucumber on the side. Thus, the Korean poet Mok-Wol Pak wrote "The Wanderer," stating "twilight over a riverside village/where wine is mellow/the wanderer goes/as the moon in the clouds."

There are five virtues in Makgeolli. The first virtue is the avoidance of hunger. The second virtue is the limited intoxication degree. The third virtue was directed to keeping the body warm. The fourth and fifth virtue pointed out the energy it provided if work became too hard

[5] Makgeolli is an alcoholic beverage native to Korea. It is made from a mixture of wheat and rice, which gives it a milky, off-white color and sweetness. It is made by fermenting a mixture of boiled rice, wheat, and water and is about 6–8 % alcohol by volume.

to bear as well as giving courage to say things that were kept to oneself due to shyness.

Some fermented Korean dishes such as Kimchi[6] and traditional sauces and Makgeolli that need cold storage are full of lactic acid bacteria. Based on *Nature* and other international academic journals, the more lactic acid bacteria are stored in the body, the better obesity can be controlled as well as more helpful in decreasing internal disease. In order to increase such good bacteria, it is important to consume many vegetables and fermented products containing lactic acid bacteria, such as Greece yogurt and Makgeolli, Kimchi, and Korea's soy paste. Traditionally Koreans used to treat colds by drinking a hot bowl of Makgeolli and sweating out the infection while lying under a hot blanket.

Like many fermented Korean foods, Makgeolli is important to health, but today most of the local Makgeolli is made from imported rice to reduce the unit cost. However, some liquor companies realize the importance of native ingredients and increasingly producing Makgeolli that is made from real native rice as a premium product. Just as the Irish call their Guinness a "meal in a glass," Koreans have the saying "it is possible to live on air for half a month just by drinking Makgeolli."

4.5 Agriculture: Forming the Basis of National Existence

Agriculture is the everlasting business that feeds humanity. The food self-sufficiency rate in Japan was 78 % in 1961, but it has since been in a long decline. In the fiscal year 2006, it went below the 40 % level, declining to 39 %, which made media headlines and the topic frequently comes up in discussions. In Switzerland, whose food self-sufficiency is comparatively low compared to other countries, the rate varies between 50 and 60 %, while in Korea, Japan's next-door neighbor, it is much lower than 50 %. Korea is the country with the lowest self-supply of food out of the OECD member nations. The

[6] Kimchi is a traditional fermented Korean dish made of vegetables with a variety of seasonings.

self-supply of grain accounted for 70 % in 1980 and dropped to below 27.4 % in 2007. Apart from rice, which is still 100 % self-supplied, corn and wheat are close to being entirely imported.

One reason for the downfall of the fourth president of Egypt, Hosni Mubarak 1981–2011, was the import of wheat from Russia which accounted for 60 % of total supplies. As the growing of crops in Russia decreased due to climate conditions, it was inevitable that the export prices to the Middle East would increase. In the future due to the increasing demand for grain and unexpected climate conditions, the production will fluctuate, leading to a rise in prices.

Although Korea produces a total of 1,000,000 t of rice, 70 % of agricultural ground is concentrated on rice production. In the meantime, the world seed and grain market has become a gigantic arena of capital competition and dominates "from the seed to the table."

If agriculture is profitless, according to the logic of economy, it should be declining. The chief executive of Japan's PASONA Group Yasuyuki Nambu tells people, "DO YOU Agriculture (pronounced 'know' in Japanese)?" asking whether "Agriculture is a declining industry?" (see Fig. 4.5). He claims that agriculture is an everlasting business of feeding humanity and argues that agricultural products will become raw materials for many industries, leading to a new growth engine.

The corn farms in the Wood River area in Nebraska, USA, increased their agricultural income by 20 %, developing the agricultural industry, while surpassing the IT industry. In a project report "Agrigento Korea" by a Korean newspaper, the bond investment genius Jim Rogers claimed, "the most promising industry in the next 20–30 years will be agriculture," and Nicolas Sarkozy, the former president of France, stated, "Agriculture is a high-tech sector of the French economy, just as much as the aerospace and nanotechnology sectors."

Agriculture, which was originally the primary industry, will be transformed to the sixth industry and become the plus α industry. The first two industries (production, manufacturing) and the tertiary industry (service) will be included in the sixth industry with a convergence, in other words, engrafting "+α" which implies the fortune of the future.

Fig. 4.5 The chief executive of Japan's PASONA Group, Yasuyuki Nambu, eating an indoor grown vegetable and the company campaign, *DO YOU Agriculture (pronounced know in Japanese)?* By courtesy of ChosunMedia

In the future, as income levels rise, the consumer's perception of value will change from quantitative satisfaction to qualitative satisfaction such as well-being. Therefore, not only will people prefer what is good for their body and health but also prefer products that are produced in an eco-friendly manner. Though Korea's Cittaslow member towns are focusing on the lack of food, it is also important to prepare a good base for healthy and sustainable growth and non-gene-manipulated food, reviving the land, which is the mother of all living things.

The first aim of a Cittaslow is the improvement of the residents' quality of life. Therefore, a way needs to be found to change the declining agriculture into a cutting-edge industry. Furthermore, as an alternative way in order to convert from financial deficit to financial surplus, the number of civil servants in the local government needs to be minimized.

The Lakewood district in California, USA, with a population of 80,000 has decreased the number of public servants and has earned the nickname "the city with no public servants." Instead, many businesses in the city, such as garbage pickup and street cleaning, have been entrusted from the public to the private sector, diminishing costs and decreasing the dependence on subsidy from the central government, but instead increasing city income through local taxes. Additionally, people who are against the increase of taxes do not vote for the local head, who is wasting the city's finances through public work projects in order to gain his political advantage.

Japan has wasted enormous resources for the last 30 years, before realizing that the government-led balanced development of the country has just been a dream that cannot become true. Thus, the individual regions in Japan are continuously fighting for financial independence.

4.6 The Value of Community Business

People who cannot get the opportunity to work fall into despair. Without work it is impossible to manage human life; thus, work needs to be in the center of life. Just as work is regarded as natural, there is no developed country that has grown without creating jobs for people.

No matter which society, its ultimate resource is its labor force, its people. The reason for this is that labor force is infinitely creative. The current technology has snatched the most enjoyable feature from people, using their hands and heads in order to create and get engaged in new activities. Instead, people are left with the less enjoyable and fragmented jobs in the workplace.

The current technology enables mass production. Since for mass production purposes, the use of machines is more efficient, the human element is gradually being taken out of the operation process, and thus, creativity in society is decreasing. In the book *Small is Beautiful*, Schumacher also emphasized, "Production by the masses, rather than mass production," [85] citing Gandhi's words, "Not mass production but production by masses," in which he envisioned a decentralized, homegrown, handcrafted mode of production.

Intermediate technology to create jobs in Cittaslow. The policy in Cittaslow areas puts priority on small businesses rather than large-scale businesses such as manufacturing and farming, rural-oriented businesses rather than urban-oriented businesses, and labor-intensive businesses rather than capital-intensive businesses, thus enabling them to escape from a technology-dependent lifestyle. Therefore, it creates medium technology suitable for small regions such as Cittaslow member towns.

Essentially, mass-producing technology is violent. It destroyed the ecosystem and humanity by wasting nonrecyclable resources. The intermediate technology[7] that caught Schumacher's attention during his stay in Burma [86] has its origin in Buddha's value of moderation.[8]

Appropriate technology does not destroy humans but instead is a means to serve human needs. The "medium technology" is far more advanced in terms of productivity compared to traditional technology. On the contrary, it is much cheaper and simpler than the capital-intensive Western technologies. Therefore, the products created with the medium technology have a higher competitive edge than items manufactured using Western technologies.

Going the road of medium technology does not imply to return to old systems. Although it might be slower, this technology seeks to create an economy with a human face (humane economy) and depends on the power of warm technology. This kind of environment has already been implemented in traditional industries and improving them with the suitable advanced know-how, using raw materials for production and creating jobs for locals. Thus, medium technology refers to the discovery and utilization of on-site or local raw material. Now we see non-sustainable growth where there is no increase in employment; therefore, the way to allow for the increasing birthrate is to creating jobs for locals through medium technology, promoting childbirth, encouraging immigration, and attracting visitors.

---

[7] Appropriate technology is an ideological movement (and its manifestations) originally articulated as "intermediate technology" by the economist Dr. Ernst Friedrich "Fritz" Schumacher in his influential work, *Small is Beautiful*.

[8] The Middle Way or Middle Path is the term that Siddhartha Gautama used to describe the character of the path he discovered that leads to liberation. The Buddha describes the Middle Way as a path of moderation, between the extremes of sensual indulgence and self-mortification.

Mirae Tech is a small but strong company in Hanam City, Gyeongsangnam-do, Korea, that provides thousands of jobs for young people. This company produces wind power-generated "tower flanges" that are adjusted to the local characteristics and provides a competitive starting salary for high school graduates as well as university graduates who are not afraid of new changes and challenges. As such, there is a hope in regional businesses using intermediate technology, as well as in sustainable agricultural businesses. The miserable unemployment that occurs in rural areas triggers a massive shift of people toward urban areas [13]. In order to resolve the problem of centralization of the population to the cities, urban businesses should be capital intensive and need to be labor economical, while rural business should be labor intensive and capital economical.

Medium technology does not rely on complex facilities, but can still provide a considerable level of productivity. Therefore, it is regarded as the suitable method for Cittaslow areas. In the case of harvesting crops, if using the scythe refers to level 1 and the utilization of combined machinery level 10, let's think about what the medium level could be.

For the development of intermediate agriculture in Cittaslow, farming needs to be independent of things, such as heavy machinery, energy intensive technology, and chemical farming. This establishes a safety net where agriculture will require mass labor forces in cases of emergency. The following describes a method to save the medium technology from becoming extinct [13]:

Small
Simple
Low-capital cost (reduce capital intensity)
Nonviolent technology (reduce technology-dependent lifestyle)

Surely, the statement "We cannot stop development. We cannot go against the general trend" is a much easier reaction compared to keeping a positive perspective. However, we cannot ignore the question "what is development?" The only thing left to decide is "how to develop faster, but comfortably and easily" and "how to develop in a healthy and sustainable way."

It is doubtful that the slow progress of technology can be a green light to the growth of the labor market. Despite generally increasing growth, undereducated workers have lost jobs due to up-to-date equipment, that is, the past decades' superspeed advanced technology has worsened unemployment and inequality. Paradoxically, technological advances have put pressure on humans, alienating them from a happy life. Now it is time to look back at slow technological progress and slow growth and enjoy its benefits.

Sustainable community. The three defining features of Cittaslow are nature (green), identity (traditional culture), and community (harmony) [87]. Poundbury, the human new town in the southwest of England, has a strong sense of community. The development officer, Simon Kearniber, claimed that the most important factor in vitalizing a village was the "density" and stated that each hectare should incorporate around 40–50 people. The current density in English suburbs in the urban area lies at 20 people, which includes 40 % of the total retired senior citizens in England [88].

In order to improve the sense of community, the Pommery Square was established in the center of town as well as pubs and restaurants in the surrounding areas to facilitate the social aspects of the community. The town is designed to support and encourage people to walk and to use bicycles, and apart from the main road, it is quite hard to find any straight roads. Such winding alleys are what highlight humanity. No matter if going to school or to the shopping district, everything is located within a 10 min walking distance, while cars need to stop and go through the winding passageways. The eco-chocolate factory is located in the center of the town creating a close work-living condition for many residents. The traditional village and buildings underline the warmth of the humane community.

Aylsham, one of the Cittaslow member towns in England, is a remarkable place in terms of practicing community culture. Figure 4.6 illustrates eight criteria of a sustainable community. It contains the desire for sustainable life through respect for the quality of life while sharing potential solutions for tackling common community issues. The eight listed criteria are currently attracting international attention due to its beauty and harmony, which embraces consilience and synthesis [89].

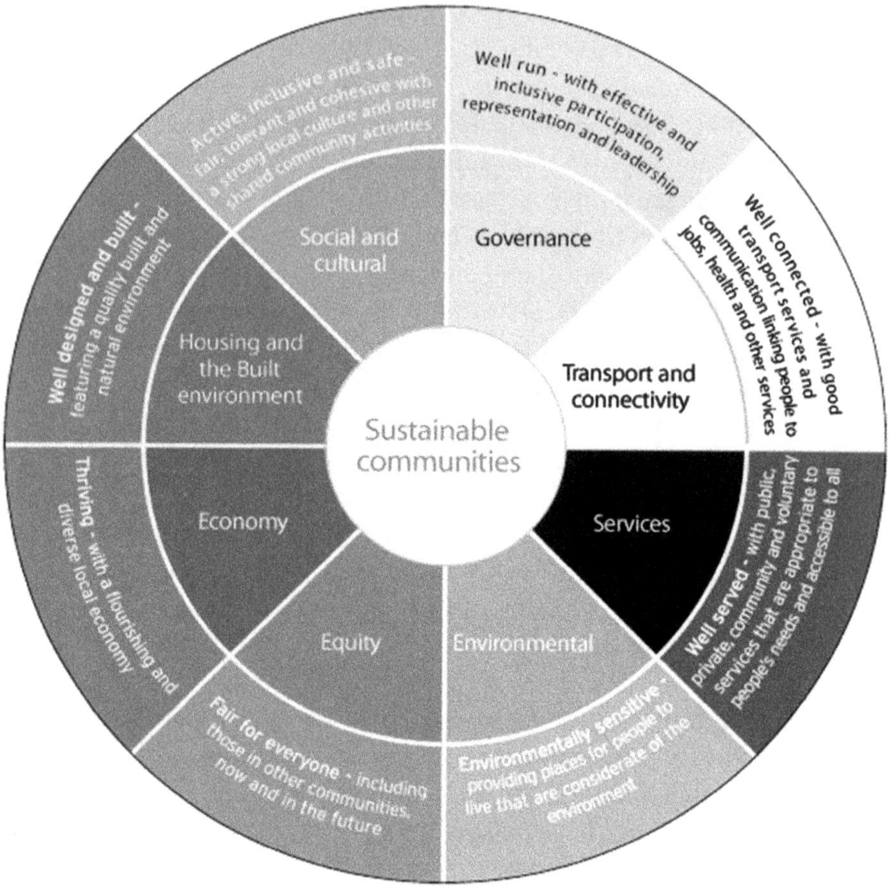

Fig. 4.6 Eight community areas aimed for sustainability by Cittaslow Aylsham in England

Even compared to world history, the Korean community culture has a long history and tradition. The philosophy of "Hongik (devotion to the welfare of mankind)"[9] was employed as the founding philosophy of the country. Also, it is a unique community culture based on the following: nature, art, and the appreciation of life's harmony between heaven, Earth, and man, while the interest of the community and family is prioritized over the individuals' interest.

[9] "Hongik" means "man's life should benefit all of humanity," "to live and work for the benefit of all mankind," or "benefiting the world through love and service."

Fig. 4.7 The United Nations declared 2012 as the International Year of Cooperatives (*Source*: social.un.org)

There is a saying, "I feel bad if my cousin buys land." Although this statement now usually refers to the envy and jealousy, it originally described the effort to produce more dung to help cultivate the land thus connecting it to the loving relationship and comradeship between relatives. During the Japanese colonial era, this statement was cleverly modified by the Japanese to point out the jealousy of Koreans, resulting in many Koreans only being aware of the distorted meaning. In Korea, all cousins were regarded as siblings. If looking at "Yesan," one of the Cittaslow member towns in Korea, the name of the city represents the place where brothers who have a good relationship reside.

The new model of capitalism – the cooperatives (see Fig. 4.7). To save local economies vital to the national economy, money and human power are important, but both of them are centralized to the central government or big cities. Provence, 1 of 22 districts in France, is depicted as a hometown in the heart the people, where local residents enjoy happiness strolling among olive trees and having wine and meals made from in-season ingredients.

According to the founder of the Cittaslow movement, Paolo Saturnini, if we are aware of our roots, our specialty, and the things that we are good at, the local economy will be revitalized despite our

Fig. 4.8 Paolo Saturnini, who inspired the foundation of the Cittaslow movement. By courtesy of the National Cittaslow Corea Network

slow life. Even by living slow, the area will be competitive and prosper (see Fig. 4.8).

"Community business" is a newly coined term. It refers to a business that is based on local resources and tackles the problems of the community through the strength of its residents while acting from a business perspective [90]. Internationally, some towns and cities do not only focus on a business to serve the public interest but also to generate profit and vitalize the region.

While global businesses strive to grow bigger, farmers in European countries such as the Netherlands, Denmark, France, etc., endeavor to survive in the competitive market. Their ultimate competitive advantage comes from their cooperative association. Capitalism that supports monopolization instead of shared profit threatens the existence of a market economy. On the contrary, cooperative association, which divides the profit equally among associates, emerged as a new model of capitalism, and Europe is regarded as heaven of cooperative associations.

Let's take a closer look at the Korean example. Since 1989 the chili paste-producing factory in Sunchang County in Jeollanam-do, Korea, has increased its profits from zero to USD 277 million within 20 years. However, it only created 357 jobs and purchased chili and beans locally produced in the Sunchang area worth USD 1.4 million in the form of "contract farming" [91]. At the same time, more and more people have been leaving the Sunchang region, resulting into a reduction in population from 49,000 to 30,000. Although the Sunchang chili paste has succeeded very well, the local economy in Sunchang could not benefit from its success.

4.7 What Is the Farmers' Market?

The world is unifying to form one community in an unimagined way, and the sense of community is becoming much more important. However, in neoliberal markets, such as the FTA, the global warfare has already begun, leaving the ones without a competitive edge to perish helplessly.

Such a thing as the number 2 does not exist in the market economy, and its effect is even more severe in the global competition. In order to escape from such a ridiculous economy structure and avoid discrimination and uncertainty, one needs to quickly adjust to the ever-changing "choice and focus" trends. There is not only one answer in the Cittaslow movement. It does not involve any costs to change one's opinion and mind.

Traditional markets have a vital position in the national economy, especially in rural areas, but the traditional market is under threat. Places that had always been the trading center of the community are currently being driven out of business by modern markets like malls, super-malls, and supermarkets. Goods sold in traditional markets are almost all available in the modern market and offered for a cheaper price. According to the Market Management Promotion Agency and the Korean Chain Store Association, Korea has closed 178 traditional shops within 7 years and has instead opened 695 business supermarkets since 2003. However, especially in a global era like this, we need glocalization (globalization + localization). The global world is not bumpy but even, and people speed over the paved highway. It is like a jungle where the weak fall prey to the strong.

Glocalization encourages the global mind-set, while keeping the bumpy local culture and adopting global advantages. Examples of successfully transforming Slow Food into an international brand are Spain with its paella from Valencia, the Cittaslow Abbiategrasso in Italy famous for its Gorgonzola cheese, and the Italian Cittaslow Greve in Chianti for its famous wine Chianti Classico. Those products gaining popularity all over the world have transformed small unnoticeable local towns into international names. If we want to create international names out of our products, we should also consider ways to give incentives to farmers who develop local specialties and cultivate high-yielding food.

The Italian blue cheese Gorgonzola (see Fig. 4.9) has a long history of fans outside of Italy – especially in France, Switzerland, and Great Britain. Today Gorgonzola ranks third in Italy in terms of cheese production, after Grana Padano and Parmigiano-Reggiano – the hard, salty cheeses popularly grated on pasta dishes. In Taepyeong Salt

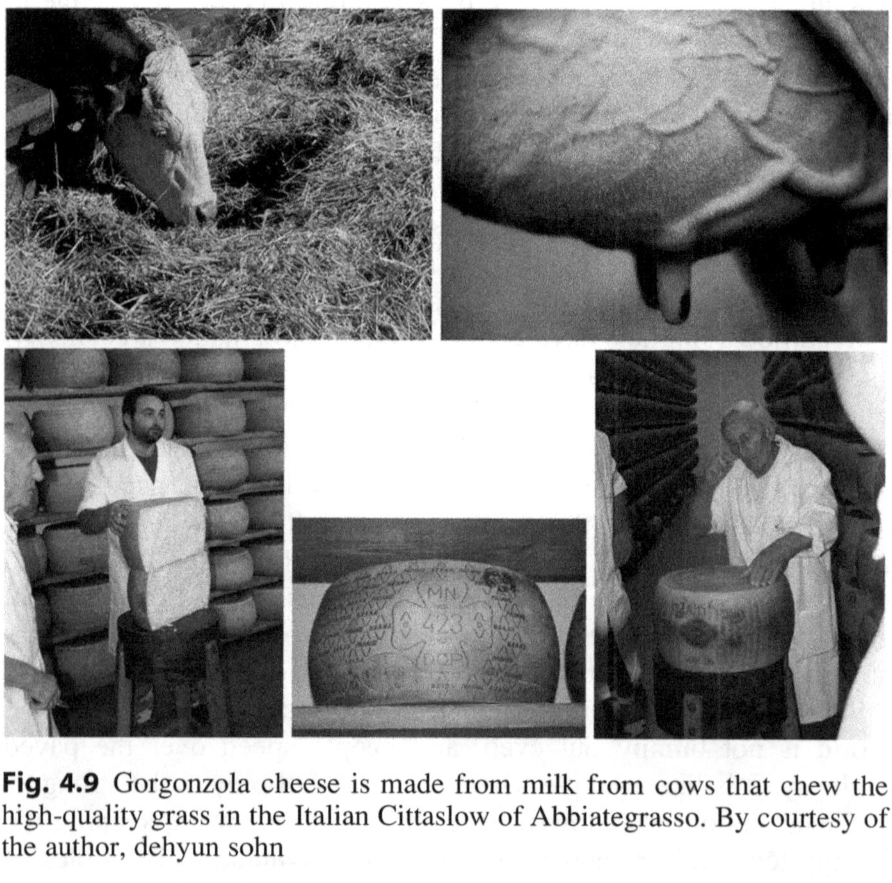

Fig. 4.9 Gorgonzola cheese is made from milk from cows that chew the high-quality grass in the Italian Cittaslow of Abbiategrasso. By courtesy of the author, dehyun sohn

Farm located in Sinan, one of the Korea's Cittaslow member towns, the regional product "bay salt" originates from the mud flats. The salt full of 88 kinds of minerals helps blood circulation [92] and is exported to Japan. Another Korean Cittaslow Sangju developed dried persimmon chocolate, which was approved as space food by the Institute of Biomedical Problems of the Russian Academy of Sciences (IBMP) at the State Scientific Center of the Russian Federation.

The farmers' market, the flower of community business. There is an enjoyable market place in Orvieto, Italy [93]. In the front place of the market, you can find food courts, kitchen utensils, and all kinds of daily necessities. Since people buy necessities twice a week on the market, there is no need to go for a trunk full of shopping; thus, food in the refrigerator is rarely decaying. Among the external merchants coming with trucks filled with products, local residents are selling vegetables and fruits grown on their gardens. One elderly woman with an old scale in her hand says, "I have sold my products here for 40 years and raised my children."

Local residents gather around the Duomo di Milano (Milan Cathedral), occupied by tourists during the day, as soon as the day darkens. This is the traditional walk "La Passeggiata." Couples pushing a stroller, the elderly couple with the grandchild, and youngsters who meet for a drink in a coffee shop on the way home to chat, everyone is busy pointing and greeting each other until 8 p.m., when people leave all together to have dinner at home. This sight can be observed on the Main Street that is lit only by the neon green light from the pharmacy sign and street lamps that are turned down making it possible to enjoy the stars at the same time.

The representative traditional farmers' market, the Spitalfields market, originated in England in 1682, attracts around 600,000 tourists and locals every weekend because of its 330 years of proud history. Besides the many products, one of the highlights chosen by suppliers as well as consumers in the Spitalfields market is the elegant glass roof and the park just next to the market where people can relax and enjoy sunbathing in summer. Additionally, one can find many fashion shows and cooking competitions taking place at the same time [94].

In Vancouver, the top destination to live, many foreigners make use of the Community Market in the City Center. This market prides itself on its core values and quality standards of producers and products.

Seven Core Values

1. Organic
   We sell the highest quality of natural and organic product available.
2. Delight
   We satisfy and delight our customer.
3. Happiness
   We support team member happiness and excellence.
4. Education
   We promote the health of our stockholders through healthy eating education.
5. Prosperity
   We create wealth through profit and prosperity.
6. Environment
   We care about our communities and our environment.
7. Partnership
   We create ongoing win-win partnerships with our suppliers.

Six Quality Standards

1. Trust
   We carefully evaluate each and every product we sell.
2. Pure food
   We feature foods that are free of artificial preservatives, colors, flavors, sweeteners, and hydrogenated fats.
3. Taste
   We are passionate about great tasting food and the pleasure of sharing it with others.
4. Quality
   We are committed to foods that are fresh, wholesome, and safe to eat.

5. Organic
   We seek out and promote organically grown foods.
6. Well-being
   We provide food and nutritional products that support health and well-being.

A traditional market and "slow money" revive a community. The Korean village market traditionally operates for 5 days. It is considered the slowest area in Korea and represents the slow life where Makgeolli as well as other fermented products and healthy ingredients (eco-food) can be found. In this market, one can experience the quality, joy, and happiness of life.

Korea is on the verge of passing a law to regulate the establishment of business supermarkets (SSM: supper supermarket) within 1 km in the traditional street market. The slow money (compared to Wall Street's fast money, it stands for money that is only valid in the specific local community) that is invested into local shops and supports the local economy should be used to create jobs for locals and encourage self-sufficiency of the local area [95].

The fast money from Wall Street, which originated from the flower of capitalism and turned into a "casino," makes SMEs and the local economy unsustainable. The owners of small local shops reinvest their profit into the local area and also pay local tax. According to the research findings of the Self-employment Society of America, USD 1 of revenue generated from a local business equates to an add-on of USD 3 to the local economy [95]. Such movements that aim to keep the locally generated profit in the area also try to circulate currency, which is only valid in specific local communities.

The farmers' market that sells locally produced products has increased threefold within the last 3 years resulting in 33 times more people since the 1990s that prepay farms to harvest products and reserve them. While standardized manufacturing goods are sold on Internet shopping malls or discount stores, we should buy locally produced organic ingredients and handicrafted products for sustainable richness (Fig. 4.10).

Fig. 4.10 The fast money from Wall Street (*Source*: www.flickr.com)

4.8 Points

In Cittaslow, agriculture is a slow business that has to be saved. Agriculture is a slow industry representing tradition and life, which is the most fundamental business. There is a sky overhead and earth under foot around the society, where it will be happy to live a life while making something, such as agricultural products. It does not cost anything to save the healthy and beautiful rural areas. The problem does not lie in the economic poverty, but in the poverty of thought.

Nowadays, farms seem to disappear in masses; however, the major problem lies in the vanishing native breed. The variety of breed has been taken over by multinational companies. In the twenty-first century of food war, especially in the war of seeds, many countries have already lost their sovereignty. In order to maintain the safety of food, we need to consider establishing a union of Cittaslow member towns to collectively manage the native breed.

The foot-and-mouth disease is a sad tragedy that was caused by the greed of humans to produce livestock faster in order to serve the mass consumption. The Cittaslow encourages a slow and deep ecology and recommends local eco-food. Healthy eating and a proper way of consuming food are the ways to protect our environment.

Between heaven and Earth, the best food is the bean. This is why it is called "the happy bean." The main ingredient of soy sauce, soy paste, and chili paste is the soy, and Korea was once called the "soy nation." Agriculture to eternally feed humanity cannot become a declining industry. "Agriculture forms the basis of national existence," a true statement for the present as well as the future. We have to preserve our sprits and wealth of the future.

The "medium technology" by Schumacher has a value that needs to be considered by Cittaslow areas. Although it might be slower, this technology seeks to create an economy with a human face and depends on the power of warm technology. In order for Cittaslow to be successful, community businesses such as the farmers' market, traditional markets, and slow money are necessary to accomplish the localization and the local community. In the economy and business world, which is dominated by the one who adjusts the fastest and takes everything, the mutual and solidarity-based cooperative businesses gain attention as a new model of capitalism.

Capitalism, Manbonism, and Happiness 5

Contents

5.1 The Money-Obsessed Contemporary Humans

According to the United Nations Human Development Index 2011, Norway was declared the world's happiest country among 187 countries, followed by Australia, the Netherlands, the USA, New Zealand, Canada, Ireland, Liechtenstein, Germany, and Sweden. The index measures happiness according to factors such as income, education, health, life expectancy, economy, gender equality, and sustainability. Bhutan, the only country that gauges its progress by measuring its people's happiness rather than economic contributions, came a rather lowly 141 in the Index. Just missing out on a top 10 spot was Switzerland in 11th place. The USA came fourth, while the UK came 28th, below Hong Kong (13th), Korea (15th), France (20th), Singapore (26th), and the Czech Republic (27th).

According to the Gallup World Poll of more than 136,000 people in 132 countries from 2005 to 2006 regarding questions about happiness and income, the results reveal that while life satisfaction usually rises

d. sohn et al., *Go Slow and Curvy*,
DOI 10.1007/978-3-319-14854-0_5

with income, happiness doesn't. Led by the University of Illinois professor emeritus of psychology Ed Diener, the global study of happiness shows that life satisfaction and enjoyment of life are two components of happiness. Life satisfaction is more closely associated with income, while positive sensibilities also depend on other factors, such as the sensibility to be respected and connected to others. "Everybody has been looking at only life satisfaction and income," he said. "And while it is true that getting richer will make you more satisfied with your life, it may not have the big impact we thought on enjoying life." For life satisfaction, Demark shows the highest score of 7.8, followed by Canada (7.7) and Norway (7.6). Korea ranks in the middle at 6.1, low compared to the average OECD (6.7). Brazil (6.8), Germany (6.7), and Italy (6.4) are similar to each other, while Hungary, Portugal, Turkey, and Greece have a relatively low level of overall life satisfaction, with average scores of less than 5.5.

Dropping happiness as the economy grows. Will the world be happier without GDP? Although Bhutan's economy is still based on agriculture and has a relatively low GDP, the Bhutanese people rank among the top 20 happiest in the world. This contrasts with the USA, which delivers much less happiness per unit of GDP. A survey conducted by Gallup Korea shows that between 1992 and 2010 the GDP per head tripled, whereas 10 % of people experienced less happiness.

Now, people in the world have the right to be happy, and recently the Index of Sustainable Economic Welfare (ISEW) is receiving attention as an economic indicator to replace the gross domestic product. Rather than simply adding together all expenditures like the gross domestic product, consumer expenditure is balanced by such factors as income distribution and costs associated with pollution and other unsustainable costs.

The homo economicus is regarded as an economic human being that exists for production and consumption, and today's materialistic lifestyle accelerates the intense greed driving the modern economy. This economy has already conquered almost the entire world.

The small country Bhutan in the Himalayas has introduced the gross national happiness (GNH) index and is proud to have the highest happiness index in the world, thus being called the "country of happiness." However, even Bhutan's happiness has stopped with its development. The King Jigme Khesar Namgyel Wangchuck who ascended the throne after his father in 2006 at 26 years of age claimed, "Without a steady economy, it is pointless to mention happiness" and has taken the lead in the country's development. However, the Bhutan people have decided against this development strategy due to the concern for the threat to their traditional lifestyle by giving the example of *The Progress Paradox* which they have seen in other developed countries.

GDP, the index which was created in the twentieth century, remains the "index king" in the twenty-first century. Some other countries in the world are also still stuck in the myth of "the higher the GDP, the better." They are filled with optimism "that if the GDP increases, they will be one step nearer to happiness than before." Some presidential candidates still promise to revive their national economy putting development as their core policy rather than promising to improve the happiness of the population. Therefore, the business focus, and market triumphant culture, has dragged the community's integrating function into chaos. A recent study by the Korea Chamber of Commerce and Industry stated that the Korean CEOs are not happy; they have an average happiness index of 78 points. This is due to the growing intensity from fierce competition and labor-manage dispute, while the business is getting bigger.

As economic growth is the priority in today's politics, the people are also tirelessly prioritizing money, making comparisons, and competing while being constantly on a treadmill without the time to be happy or even to take a breath. The Gallup Korea partner, Global Market Insight, took a survey of 5,209 adults in ten countries including Canada, China, France, Germany, Italy, Japan, Russia, Korea, the UK, and the USA on December 2010. It showed that Koreans are fatigued by a persistent yearning for material wealth, and a mere 7.2 % of Koreans feel money and happiness are not related, compared to the average of 24.3 % in the surveyed countries. Most people living in happy countries according to the survey report themselves as being "probably the happiest person they know" (33.9 %). Brazilians are the happiest people in the world with 60 % saying that they are happy, followed by Vietnam (49.1 %), the USA (29 %), Canada (27.7 %), and Malaysia (26.8 %). Interestingly 49.3 % of Koreans think Bill Gates is the happiest person, while the Dalai Lama is favored in some countries of higher economic standard such as Denmark, Canada, Finland, etc. Also, more than half of respondents consider money to be the main reason behind Korea's low birthrate.

In 4 BC Aristotle mentioned in his speech, "A state's purpose is not merely to provide a living but to make a life that is good, and the purpose of a government is to enable its citizens to live a full and happy life." This nation, Greece, is now facing a national bankruptcy crisis. Since the government has cut public servants' salary 20 % to regenerate the national economy, it has encountered intense protests.

To live happily is a common topic in humanity. The ancient Athenians said, "Humans have to live well," the ancient Rome showed its significant aesthetics during the Renaissance as well as powerful military strength, and the Age of Victoria in England tried to put the values of life on cultural honor and sought quality of life during the period of Industrial Revolution when money was king. In other words, those pioneers did not define the key to happiness as money, but as the aesthetic consciousness, honor, and freedom. Cittaslow should also focus on generating a city, where people can relax and increase the quality of life instead of economic growth.

According to the abovementioned survey by Gallup Korea partner Global Market Insight, Korea is suffering due to work, while Finland is happy due to work. 63 % of Korean respondents claimed to look at salary and job security when selecting a job, whereas in Finland 60 % claimed that the type of work was more important. This showed the difference in cultural level as well as level of happiness.

According to John Keynes, British economist, the current economy respects the means to success rather than the purpose. The economic development should not be regarded as the goal, but the priority should be the happiness of the people. According to the result of Professor Ed Diener's research, if an income of USD 10,000 would suffice to tackle the problems of basic needs, then a higher income would put happiness as a threshold value into inverse proportion [96]. The British economist Richard Layard wrote in his book *Happiness: Lessons from a New Science*, "once income per head exceed about USD 20,000 (at 2005 prices), extra income appears to have very little additional impact on happiness." The Bible says, "For the desire of money is the root of all evils; which some coveting have erred from the faith, and have entangled themselves in many sorrows (The First Epistle of St. Paul to timothy 6:10)."

Greedy and intense economic development policies such as the "Miracle of Han River" over the last 60 years are now hitting a dead end. People say that you need long sleeves to dance and a lot of money to run a good business. For politics, however, we need to take a different approach.

Economics is not an independent subject, and it has originated from meta-economics. Meta-economics refers to a new kind of economics. It describes the progress from quantitative growth to qualitative growth and qualitative sense, making the qualitative difference to reach the stage of an advanced country. It is not about the survival, but about the standard and quality of life. The quantitative aspect is secondary to the qualitative aspects.

The work of "warm capitalism." The French author Andre Gide (1869–1951) said, "In order for me to be happy, the happiness of everyone is required." Eric Weiner, an American columnist, described in his book *The Geography of Bliss: One Grump's Search for the Happiest Places in the World* that the "strength of community" creates a happy world [97].

The 2010 Gallup World Poll on The World's Happiest Countries revealed that Denmark was ranked 1st, followed by Finland, Norway, the Netherlands, Costa Rica, and Canada, whereas Brazil ranked 14th, Mexico 18th, Argentina 30th, and Korea 56th. The happiness indicators of Denmark were based on the mutual acceptance and respect in society. From childhood onward, the first thing people had learned at home and at school was to respect others and not to trouble the next person. Respecting others implies that corruption or unjustified benefits are not accepted and form a strong ring of trust among the members of society.

Furthermore, Denmark possesses a very "transparent government" system. The thought of "getting as much return as I paid for" rather than "tax being pulled out" is stuck in the roots of peoples' minds. In the Transparency International study of 2007, Denmark ranked first among 180 nations in terms of corruption perception index (CPI), which implies that public servants as well as politicians in Denmark are the cleanest, with perennial highfliers Finland and New Zealand.

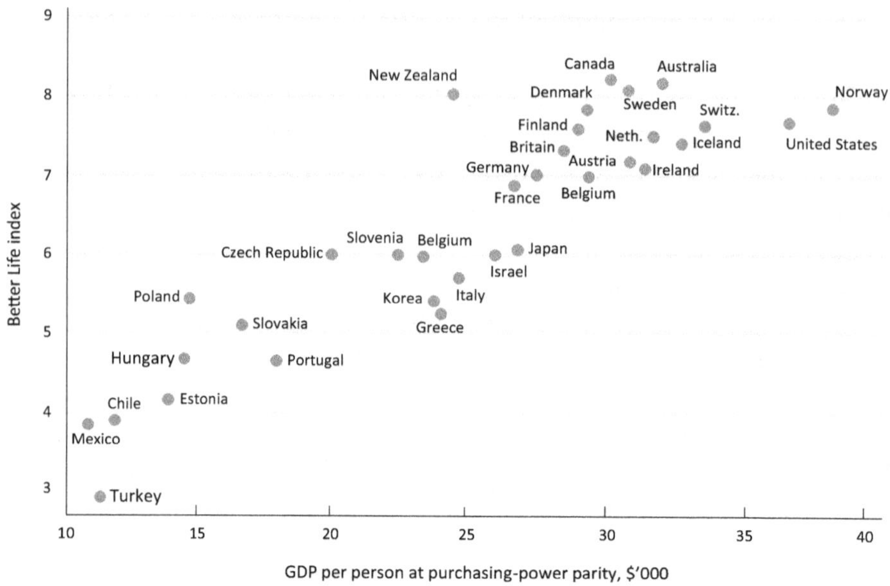

Fig. 5.1 OECD Better Life Index (10 = best) and GDP per person (*Source*: OECD, 2009)

Drivers are stressed on a daily basis due to the rising numbers of cars. As the GDP increases, more petrol is being utilized; this cannot actually act as economy index, just as it cannot represent an index for the happiness of the people. Australia topped among 34 OECD member countries in the Better Life Index of 2011 followed by Canada, Sweden, New Zealand, the USA, and Norway (see Fig. 5.1). The index is compiled based on 11 categories such as housing, jobs, income, and education. Korea ranked 26th as only 36 % of Koreans are satisfied with their current life, and Korea's neighbor Japan was placed 19th. Korea scored high in education, safety, and jobs but was in the bottom group in work-life balance, housing, and income. The OECD average for life satisfaction was 59 %. Northern European countries ranked high in the category, with 91 % in the Netherlands, 90 % in Denmark, and 86 % in Finland saying they like their life. In Japan, only 40 % said they are happy. The weakest category was identified as "community life" and was six points below the average score of developed countries.

Some Asian societies like China, Japan, and Korea that were built traditionally on the values of family and relationship are currently failing as these relationships weaken. The Swiss born Margaret Ningetto, who has served as a volunteer in Korea for many years, said, "When I first came to Korea, there were many poor people. However, there were also many happy people. The Korea today might be richer than in the past, but I think there are more insecure people. Previously, people helped each other out even with just a spoon of rice, but nowadays, everyone is only thinking about themselves and their family."

Kids grow up hearing in America "you should be of assistance to a person," in Japan "you should live without becoming a burden to others," and in Korea "you should not lose out to others." There are still allegations of corruption in today's society and no morality of leadership in the world. Such leadership of society will only create conflict, trouble, and mistrust, while happiness steadily disappears and living conditions become more difficult with people getting depressed and social problems increasing. The psychology professor Jonathan Haidt from Virginia University, USA, said, "Humans are social animals. Happiness cannot be found alone, nor fulfilled, but the happiness of social animals is to be found in the relationship with one's family, one's friends, and one's workplace." That is, people who focus on achieving wealth and career advancement are less happy than those who focus on good works, religion or spirituality, or friends and family.

According to a study of Kahneman and Deaton (2010) [98], who asked 450,000 respondents to report on both emotional well-being and life evaluation in 2008 and 2009, "people's life evaluations rise steadily with income. Importantly, the same percentage increase in income has the same effect on evaluation for everyone, rich or poor alike even though the absolute dollar amounts differ. For example, a ten percent increase in income moves everyone up the same number of rungs, so someone earning \$20,000 a year who experiences a \$2,000 increase would move up the rung at the same rate as someone earning \$100,000 a year who experiences a \$10,000 increase." 40 % of Koreans think that money is the main criteria for happiness. However, surprisingly, until one earns a certain sum of money a month, the happiness index is sharply increasing, but decreasing once it goes over this limit. These results come from the careless relationships with the family, friends, and other people in order to earn that money.

In the past, the base among business people was built on courtesy and trust and thus formed a warm capitalism through relationships. If we take a look at a painting, which shows the Deoksan market in the Korean Cittaslow Yesan (see Fig. 5.2), we can also see relationships encouraging courtesy and virtue. In the fifteenth century, this kind of relationship was widely spread within Korea.

The life of ordinary people has always been around the marketplace. Although, bartering was the primitive form of commerce, organizations were formed by peddlers, and they created the full-scale market around them. In Korea, those markets appeared in the fifteenth century, and by the mid-eighteenth century, 1,064 markets were created in the whole country.

These marketplaces were not only an area where people could trade items but also the place to receive gossip and news about the outside world, where people met and acknowledged each other and enjoyed social interactions. It was also the place where performers, medicine peddlers, singers, and other street beggars could gather and share their talents and entertain others. Marketplaces have historically been a place for social entertainment. As such, the marketplace was the room where commoners could share their emotions, such as joys and sorrows, and open to all kinds of topics.

There is another picture which shows the family relationship quite well (see Fig. 5.3). This creation, called "The Family Leaving the Road" by the Korean genius painter Joong-Sub Lee (1916–1956), describes the families torn apart due to the war and their sad longing for each other. The picture was originally meant to show the family on an ox cart going to the warm South and was included in a letter of the separated family. However, the picture that was created through the times of poverty was sold on an auction for USD 1.8 million after the painter passed away alone, which is quite ironic. Let us assume that this picture conveys happiness to people through the relationship of the family.

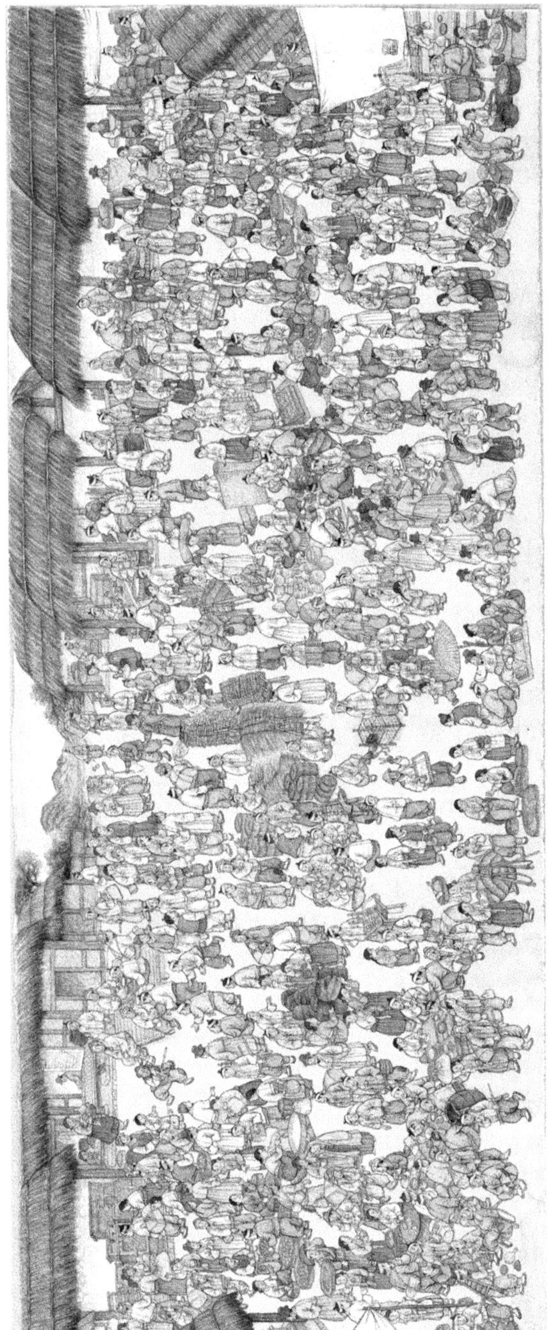

Fig. 5.2 "Marketplace Scene" that describes an eighteenth century marketplace of a Korean housed in the Peddler Relics Exhibition Hall in Cittaslow Yesan, Korea. By courtesy of ChosunMedia

Fig. 5.3 The family leaving the road (1950s): the Korean painter Joong-Sub Lee (1916–1956). By courtesy of Korea Database Agency

5.2 The Soul's Fundamental Is "Slow"

Happiness is not gained, but happens. Life is significant. However, even this precious life will someday end and perish. Everything that is done by humans in their lives is all in return for happiness. Expert skills of chefs, authors, and other artists are making others happy through their work.

It might be due to this phenomenon that the values of capitalism are also currently changing, through the classic laissez-faire capitalism (Capitalism 1.0), the modified capitalism through the influence of the government in the 1930s (Capitalism 2.0), the neoliberalism that was emphasized by the autonomy of the market in the 1970s (Capitalism 3.0), which all encouraged the self-regulating function of the market and the social responsibility of those participating in the market. The author of *Capitalism 4.0* English economist Anatole Kaletsky emphasizes "warm capitalism" with the aim of a happy joint development; that is, seeking to expand the pie of economic prosperity for everyone through market-centric means requiring social consensus.

Echo-boomers or post-baby boomers in the world are suffering from economic problems such as unemployment, which result in the decrease of marriage and childbirth. Therefore, they are called "Screwed generation (or Generation Y)" in the USA, "Baringheowoo generation"[1] in China, "Dankai juniors"[2] in Japan, and "880 thousand-won[3] generation"[4] in Korea. People demonstrate with the motto "Occupy Wall Street" against the distorted distribution of wealth on highly twisted 1/99 basis; in such an atmosphere of anger, it is rather easy to talk about warm capitalism, but much harder to fulfill it.

According to the "Future Korea Report" on the quality of society of OECD member nations by the Society Development Research Institute of the Seoul National University in Korea (2011) [99], Denmark ranked first and Korea 28th as the lowest rank. The striking point was that Danish youngsters were drastically open to try out creative types of work, whereas Korean youngsters were busy avoiding any risk factors. It is important to make youngsters say, "I succeeded as I was doing what I enjoyed, and thus I am happy" in order to build a bright future for the nation.

In a survey of "100 years, a blessing or a disaster?" it was found that seniors who earned even small amounts of money every month were twice as happy as seniors who did not work at all. Senior citizen stated that they preferred to work although their spending money was sufficient. The Korean poet Shi-Hwa Ryu stated in "All living things should be happy," "No matter how poor, if you have a heart there is something to share. If you share the resourceful mind, material things will follow naturally like a shadow" [100].

Boulder in Colorado was pointed out as the happiest city in the USA. The city emphasized the employment of local shops rather than cheap chain stores. Additionally, there are 241 km of bicycle routes; thus, 20 % of the business people use the bike to and from work, and 1 out of 3 have a dog at home. Furthermore, when a policy decision is to be made, the citizens are included in the decision-making process. It is the law to include citizens whenever three members of the city council gather for a meeting. This gave birth to a happy community.

---

[1] Baringheowoo means "after 80" and refers to those born after the economic reform the era of the 1980s.

[2] Dankai juniors are children of Dankai generation who are baby boomers born after the Second World War.

[3] The won is the monetary unit of Korea.

[4] Their average monthly salary ends up being 880,000 won (USD 790).

Cittaslow respects the speed of nature. It is said that the bamboo blossoms only once in 100 years, even if it is planted at a different location, the biological clock of bamboo does not change. Everything in nature lives exactly the time that was given to them. Only humans are escaping from the time given by nature. Finding the original speed given by nature is to find happiness.

We were not created to live unhappily. The Bible teaches us, "What does it profit a man if he gains the whole world and loses his soul?" If we sacrifice our soul in return for wealth, what is the benefit? The Indian nuclear disarmament activist and editor of *Resurgence & Ecologist*, Satish Kumar, argues that we have sacrificed three important things in return for wealth. These are soil, soul, and society. "The future well-being of humanity and the earth depends on a new world view in which the care of the planet, nourishment of the soul and the nurturing of the human community are integrated and seen as a continuum." He adds that things, which have a connection with the earth, the connection with oneself, and the connection with other people, are considered as the origin and the solution of happiness. "I propose the trinity of soil, soul and society. Our reverence for the earth, our care of the soul and a just order in society represent a vision of sustainability, spirituality and justice."

There is a suitable teaching for the current people who have become slaves of substance to escape difficulties [85]. It is simplicity and nonviolence, which is the core of the Eightfold Path of Buddhism. The lifestyle of Buddhists is amazing and shows a surprising level of satisfaction while owning very little.

In Catholicism, there are the four cardinal virtues, wisdom, courage, temperance, and justice, which can teach the appreciation of values. Wisdom is organic, soft, nonviolent, graceful, and beautiful. Humans are clever, but they cannot survive without the wisdom of traditional lifestyle since the core of wisdom lies in sustainability and peace.

The poem "The Happiness" by the leading twentieth century Korean poet, Chi-Hwan Yu (1908–1967), ends with the phrase "Because I loved, I was really happy." There is a classic Eastern parable "If the owner of a shop gives the guests virtue through his mind, the guests will keep coming and gathering in the shop like a cloud if the mind feels comfortable in the place." The poet loved the virtue and comfort of the shop owner. In the future, goodness and making the guest feel happy will become the competitive edge. Success is therefore not the key to happiness, but happiness is the key to success.

The Happiness
by Chi-Hwan Yu (1908–1967)

To love is happier than to be loved.

Today I come to write a letter
in front of the window of the post office
where the emerald sky is shown brightly.

(ellipsis)

To love is happier than to be loved.

I write a letter to you today too.
My sweetheart, I must say good bye.
Though this is my last greeting,
because I loved, I was really happy.

Blessed are the poor in spirit. We need to pay more attention to original blessings rather than to original sin [101]. However, people do not understand the truth in the paradox of becoming a bigger sinner, the more sin is emphasized and we are suppressed by it. We need a fundamental belief that God has created the earth and humans with the promise of never giving up on them before starting to blame the world and oneself for our lives, as life was created to be good and valuable.

Jesus said that a man's life is not represented by the abundance of his possessions. This indicates that everyone is living through the mercy of God, and thus, life is a gift. It is a request "not to worry about what to eat and what to wear." Just as children are given the gift of life through their parents, we are all living freely due to the mercy of God.

2,000 years ago, people gathered at the Sea of Galilee, in the north of Israel. There, Jesus said, "Blessed are the poor in spirit, for theirs is the kingdom of heaven." This is the first blessing of the famous Eight Beatitudes from the Sermon on the Mount.

According to the Korean Catholic priest, Dong-Yeob Cha, the meaning of happiness is "happen or arise" and, therefore, refers to the word "happen" [102]. Happiness is not gained but happens. Then, we ask ourselves, who is producing happiness? It is none other but oneself. This is the one important truth and rule. Jesus enlightened our minds, so he could spread this happiness.

Such poor-minded peoples' "mind" is the spirit. How can this teaching be used in our daily life? This world is simply to be enjoyed, instead of owned. The heavenly blessings that have already been given to me, the blessings of nature, all these simply need to be enjoyed. Poor-minded people are poor in sprit. The soul is fundamentally slow, and happiness means to surrender.

5.3 The Principles of True Capitalism and the Spirit of "Dasal"

The fundamental basis of capitalism is money. The problem with capitalism does not necessarily lie in whether it is good or evil, but rather in the logic of its practice. It was a new development that political parties promoted "economic democratization"[5] bills just before the 2012 general election and presidential race in Korea, each having the same goal but proposing different solutions to reach it.

The Eastern character "economy" originated from the character meaning "administering the state to relieve the people's suffering," but it is difficult to see the authenticity of the meaning any more in today's economy. As we have endlessly followed capitalism, we have lost all kinds of morality, slowness, and human kindness. Alfred Marshall (1842–1924) who was one of the most influential economists of his times said, "Religion and economy are the two factors which have most influenced the history of mankind. If 'economy' is seeking the rich life, 'religion' is encouraging the sacred life."

[5] Economic democratization or democratization of the economy refers to an election campaign or political slogan that the nation's economy shall be managed and operated in a democratic way.

In 1517, the religious revolution had blossomed, the most important ideologies in Europe's modern history; individualism, democracy, freedom, and capitalism were being developed as well as puritanism [103]. There are three principles in puritan capitalism. The first one, "rationalization of the pursuit of profits," aims to be conservative about excessive greed, thirst for profit, and the decadent lust for money. The second one, "freedom of competition," refers to the desire for goodwill in competition to elevate the industry and reminds one that excessive competition is dirty and cowardly and leads to rampage leading to the loss of any form of peace, love, and friendship through extreme inhumanity. The third one is to "acknowledge private ownership" [104]. However, it is based on the protestant ethos that warns against those who covet riches, because they can seek unearned income and show a propensity of people who gained sudden wealth.

Today, competition is a topic all around and about. During the days of the former president George Bush, Todd Buchholz was an economic policy adviser. He stated that the human nature is competitive and does not fit into egalitarianism. Although communist countries, such as Russia, have conducted a daring experiment to remove competition and only pursue the happiness of the people, it has resulted in the mass production of alcoholics. He claimed that competition is the driving force of development and questioned whether happiness is possible without competition; he said the smartphone is also a by-product of human competition. However, we can answer the question ourselves, whether such materialism is generally providing happiness to people. We can argue that we rely on reasoning, self-control, wisdom, and justice in order to ensure we survive competition.

The basic principle of capitalism was found in the book *The Wealth of Nations* by Scottish moral philosopher and a pioneer of political economy Adam Smith (1723–1790) (see Fig. 5.4). His basic philosophy was based on the idea that in the economic activity, it is more desirable to "empower" the individual rather than to focus on the mercantilism of the nation [105]. Thus, rather than laissez-faire freedom, autonomous freedom should be considered. The reason that big businesses have invaded the passageways and replaced local shops accompanied by growth and the failure of symbiosis

Fig. 5.4 Adam Smith, the intellectual father of capitalism (*Source*: commons.wikimedia.org)

relates back to the government policy in which the government gave autonomous freedom to big companies and has left the economy in the hands of autonomy. It is a misjudgment that such big companies have a sense of responsibility based on autonomous freedom. As a result, the happiness of the people has moved further away.

Originally, ethical sentiments and morals had formed the base of Adam Smith's first capitalism, and they are found in his two books *The Theory of Moral Sentiments* and *The Wealth of Nations*. Although the latter book has been more famous than the former, in his epitaph it is written, "Here are deposited the remains of Adam Smith, author of *The Theory of Moral Sentiments*." It can be seen that he valued *The Theory of Moral Sentiments* more than *The Wealth of Nations*.

Schumacher, who seeks the Buddhist economy, said the following, "Economics is seeking of greater happiness through less consumption" [13]. He has found this truth through his experience in Burma. "After 2–3 weeks, visiting a few regions, I have understood that the people of Burma do not need the advice of an economist like me. Actually, we need to learn from them. The people of Burma have a perfect economic system to maintain the highly developed Buddhist culture."

To the question "in which ways is economy related to Buddhism," he replied, "Economics without Buddhism is like sex without love." Small things are wonderful. Everything in nature is limited by its size, speed, and strength. Nature, including mankind, is dependent on the effects of balance, modification, and purification. On the contrary, technology does not accept the fundamentals of being controlled. In this case, balance, modification, and purification have no effect.

Economy starts with the creation of demand. A lot of effort has been made to create unnecessary demand. Furthermore, demand was created through advertisements and other media. Schumacher added that people became slaves of desire and thus became unhappy.

The virtues of the spirit of "Dasal." The fourth king, Jigme Singye Wangchuck, of Bhutan, which is called the country of happiness, made a surprising decision 40 years ago. Instead of the materialistic scale of gross national product (GNP), he invented and enforced the gross national happiness (GNH) in the country.

In order to form a healthy development of the society in capitalism, neighbors and countries need to become "earning and sharing people" rather than "depriving people." Just as in the teaching of Hwaom (the Huayan school of Chinese Buddhism),[6] cooperation and harmony need to coexist for the hope of humanity [106].

In the times of the Silla Dynasty, one of ancient Korean kingdoms (57 BC–935 AD), there was a spirit of "Dasal (saving all)." Such spirit refers to the coexistence of humans, animals, and plants, which is urgently needed in today's world. If we don't turn the axis of this paradigm, humanity will face extinction in the near future.

The honorary president of Kyocera Corporation, Kazuo Inamori, who is regarded as "God of management," said, "We need to enforce the moral sense of seeking profit through honest work. As Buddha said, we need to forget about greed and learn satisfaction. Lions do not hunt if their stomach is full. Only humans keep hunting although they are not hungry anymore." The fact that big companies in Korea

---

[6] Hwaeom is the name of the Korean transmission of the Huayan school of Chinese Buddhism.

keep opening stores and bakeries in small passageways is nothing but greed. It is important to justify the diversification of a business.

Jesus once mentioned, "Do not think that I came to bring peace to earth. I did not come to bring peace but a sword" (Matthew 10:34). This statement seems contra-peaceful; however, it implies that he intended to firmly cut down all temptations and peace-disturbing elements in order to gain absolute peace. It is a message to fully cut out the human obsessions of success, judgment, material, and greed and to gain an empty heart that can be filled with peace.

Today, the unconditional pursuit of economic growth is justified in various fields of society, but it is just like the Titanic Realism where one does not care and rushes, which was established by C. Douglas Lummis who is a writer and former professor at Tsuda College in Tokyo. God's presents, "idleness," originated from the word "lazy," and "slowness," were underestimated, and their value has been concealed by capitalists, the clerics, economists, etc., who strongly espouse capitalism. Also idleness and slowness were excluded in a capitalist civilization adherent to the speed, and moreover while the puritan work ethic became powerful, labor was identified with heaven, while idleness is associated with hell. Idleness and slowness were regarded sins, which were believed to be against the divine will. "The devil finds work for idle hands." Labor was the divine will that was distorted. In the book *The Protestant Ethic and the Spirit of Capitalism* (1920), the German sociologist and political economist Max Weber (1864–1920) stated that "hard, continuous bodily or mental labour" is important as an ascetic technique and as a defense against other temptations. Among the sections of the Sermon on the Mount, Matthew 6:25–34, Jesus tells us about slowness and leisure; "Look at the birds of the air; they do not sow or reap or store away in barns, and yet your heavenly Father feeds them. Are you not much more valuable than they? Can any one of you by worrying add a single hour to your life?" Now, the question about how well we live in this splendid age of capitalism era is cast. Although the equation "fastness + competition + capitalism (money) = loss of humanity" has been completed, the horse riding dance in the song "Gangnam Style" by Korean singer Psy is a bitter satire on a capitalistic society (see Chap. 2).

According to the book *Honeybee Democracy* by world-renowned animal behaviorist Thomas Seeley, individual bees make decisions collectively and democratically with dances, and dancing is a means of communication between them. Every year, faced with choosing and traveling to a new home, honeybees undergo a process of collective fact-finding, vigorous debate, and consensus building. With the birth of a new queen, the old queen should be on her way with worker bees to seek a new hive. After worker bees evaluate potential nest sites, they advertise their discoveries to one another with dance and engage in open deliberation to choose a final site and then navigate together to their new home. More interestingly, the better the site they find, the more vigorously they dance.

John Chrysostom, archbishop of Constantinople (347–407) who was an important Early Church Father, said, "The bee is more honoured than other animals, not because it labours, but because it labours for others" [107]. Bees produce good works for their neighbor at all times by pollinating the plants for food, beauty, and air quality, which are highly necessary for the survival of others. The worker bees go around collecting nectar with which to make honey and spend wintertime together in a honeycomb, maintaining their body heat. Like bees that share food and search an optimal site for living, humans need to share good things with others, respect each other's opinions, and enter into a warm lasting relationship. Just as a queen bee restrains her own decision and lets other bees make up their own minds and make a decision, people can build an open-minded community. When bees face some changes such as selecting a living site, they prefer happiness for all to their own interest. It is the lesson that humans should learn from bees.

5.4 Greener Cittaslow^{Slowcity}

The tree, the river, and happiness. Why are we currently emphasizing "slow"? The answer is that we need to change from "development" to "maturity," "quantity of life" to "quality of life," and "speed" to "depth," which refers to the depth of thought and heart, depth in work, and depth in human relationships. The ones that are living in relation to nature believe in the communal union of animism and totemism (the symbol of family).[7] The ancients are following a spirit in animism that believes in the life of trees and rocks. If we take a look at Kevin Costner's movie "Dances with Wolves," there is a scene in which even the cutting down of a single tree needs the approval of the chief in the Indian village. There is a Native American saying, "Human can live as long as trees live." Buddha recommended the obligation of every individual to plant one tree at least every 5 years and to care for it. As long as this was observed, the whole of India would be covered with trees, free of dust, with plenty of water, shade, timber, fuel, food, medicines, and other materials.

The world famous architect Tadao Ando deplored the people of today for not thinking about the next generation and emphasized the reconstruction of lost values such as pride in the local area and the value of forests in order to improve the current situation of extinction of family and other values on the planet.

The American author, philosopher, and historian Henry David Thoreau (1817–1862) said, "We talk about paradise without mentioning nature, what other can this be, but an insult to the planet." [108] There is no need for high skills or techniques to plant and take care of a tree. Everyone, who is physically healthy, is able to contribute to it and can gain the broad scale of useful goods from trees (see Fig. 5.5).

The French photographer Yann Arthus-Bertrand had a special exhibition in Korea with the theme "Forests, People and Desertification," in which he asserts that 40 % of the planet is already a desert. On average, over the course of a year, one hectare of forest absorbs five to ten tons of carbon dioxide, while it releases between 10 and 20 t of oxygen. However, through cutting down tropical forests to cultivate crops and the random destruction of trees, global warming is accelerated resulting in frequent forest fires. Such forest destruction is one of the main causes for climate change due to the creation of greenhouse gases.

[7] Totemism is a system of belief in which each human is thought to have a spiritual connection or a kinship with another physical being, such as an animal or plant, often called a "spirit-being" or "totem." The totem is thought to interact with a given kin group or an individual and to serve as their emblem or symbol.

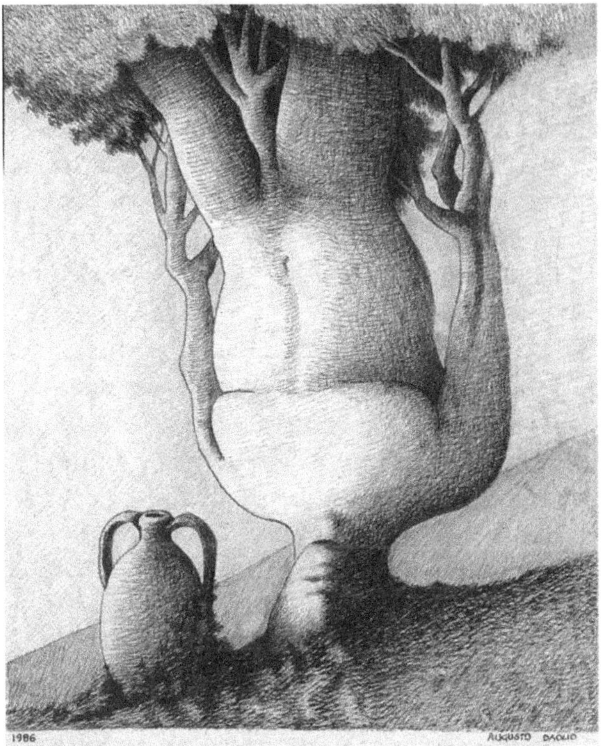

Fig. 5.5 The picture that was given to the author, Prof. dehyun sohn, for visiting the Cittaslow in Italy Comune di Novellara. Although it shows a person who is planted into the earth, if we don't pay attention, it gives the illusion of a person and a tree hanging in heaven. By courtesy of the author dehyun sohn

Despite 70 % of Japan's land consisting of mountains, 80 % of the wooded materials for construction are being imported from abroad. In Japan, the current proportion of natural forests is less than 3 % of the country. It might be due to this phenomenon that the Ayatso village in the Miyazaki region welcomes 1.2 million visitors every year that come to see the rare laurel forest. Additionally, the village hosts a food festival, where children and adults can enjoy "the tasty food education." Ayatso is famous for its organic agricultural movement on the largest scrub of 1,700 ha in Japan. This is what the village head calls, "Without a scrub, there cannot be a wide river. If river and ecosystem are destroyed, it will ruin the ocean as well."

The nuclear accidents in Fukushima in Japan in 2011 and Chernobyl in Russia 26 years ago still have significant aftereffects within the countries. The nuclear power and the nuclear power plant give the country a terrifying authority and are widely used for military purposes as well as for the industrial and economic development.

Korea, for example, is strategically using the term atomic energy instead of nuclear power in order to distract attention from its negative image. The so-called nuclear power generation or nuclear power option is accepted on the condition that there will be no natural disasters and no mistakes from human manipulation. The case for atomic power is argued within governments, media, and intellectuals as though there are no major dangers in the operation of the power plants at all. This view is comparable with the Faustian deal with the devil (see Chap. 1).

The other alternative would be the rejection of nuclear power. The overall rate of dependence on nuclear power in the globe is merely 2 %. According to the World Nuclear Association, these were the top 10 nuclear-dependent nations in 2011: France 77.7 % of electricity from nuclear power, Belgium (54.0), Slovakia (54.0), Ukraine (47.2), Hungary (43.2), Slovenia (41.7), Switzerland (40.8), Sweden (39.6), Korea (34.6), and Armenia (33.2). Germany has promised to completely avoid nuclear power by 2022 [109] and instead generate the second "Miracle on the Rhine" to produce sustainable energy. While the USA recently approved its first new nuclear reactor in 30 years, energy-hungry China has been building reactors en masse, with 26 new facilities currently under construction, another 51 in the planning stages, and 120 additional ones being proposed. The greed for energy will leave our descendants with disastrous natural environments in the future. We need to reflect on the roots of life for the reduction effect in the economic luxury and the deceptive society. Nuclear power is not a simple alternative for energy or environmental issues, but it represents a human crisis. Each individual person needs to save energy in various ways, such as sleeping with the lights turned off at night. Now is not the time to switch on the light, but to project the darkness.

In the last 100 years, the average temperature on the planet has increased by 0.74°, whereas the average temperature in Korea has increased by 1.78°. Nevertheless, Koreans are still not aware of the critical situation the world is facing through CO_2 emissions. If the temperature rises by 2°, it is said that living conditions will worsen drastically (see Fig. 5.6). Many areas in the world are continuously facing heavy rains and weather catastrophes. Korea on the other hand

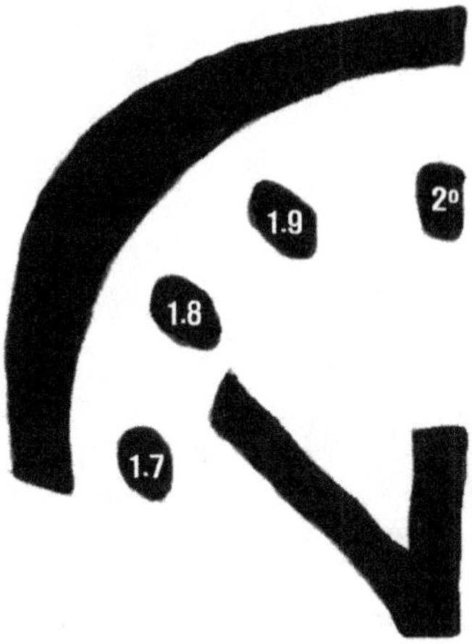

Fig. 5.6 There are 0.22° left to reach 2.00°; if the temperature rises by 2° from now, it can harm the well-being of people. By courtesy of the author, dehyun sohn

has invested USD 19 billion in 2 years for 4 major rivers projects. Through the torrent of the four rivers, they have generated a standardized ecosystem that floods away life and memories.

The river is responsible for the natural circulation of water between the ocean and the land. Every healthy river needs a flood season as well as an extreme dry season. It is not simply a path where water flows. The river forms a curvy and complex system, where mysterious life forms exist. The natural form of the river is slow and curvy, so it can support nature with its existence. God is creating the slow and curvy. The same goes for the pine trees in the mountains that are much more beautiful if they are curvy rather than straight. The river flows everyday anew. If a healthy river is destroyed, it results in the destruction of the ocean (see Fig. 5.7). Therefore, humans cannot be the owner of the ecosystem, but need to be modest participants responsible for its maintenance.

Humans are not like annual plants, but like trees that live for hundreds of years. The famous Spanish architect, Antoni Gaudí (1852–1926), left the well-known phrase, "Trees are God's best architecture. There is no better structure than the trunk of a tree or a human skeleton, structures which are at the same time functional and aesthetic." In the book *The Man Who Planted Trees*, the French author Jean Giono (1895–1970) tells the story of one shepherd's long and successful singlehanded effort to reforest a desolate valley in the foothills of the Alps in Provence throughout the first half of the twentieth century. Therefore, such a thing needs to be done through

Fig. 5.7 Naturally, the river needs to flow slowly and *curvy*. This is due to its nature. However, through all kinds of industrial development, the river is *straightened*, making it difficult for it to function properly. By courtesy of ChosunMedia

the spirit of taking care of others. The Eastern character "Mok" (tree) includes the character "In" (person) for a reason.

The tree is Noah's green ark, as well as the profit increasing green cash cow. In the twenty-first century, the real value of the tree needs to be savored. Thirawit Leetavorn, the vice president of the international copy paper company Double A, has invented an eco-friendly business system [110], which involves the seeding of fallow grounds in the banks around the rice fields of Thailand, where farmers can increase their yearly profits by cultivating the seeds. This seed production sucks in carbon emissions and thus also creates a low-carbon movement.

The German Cittaslow Waldkirch in the lap of Germany's Black Forest encourages people to plant and maintain a tree for every child that is born. They should tend and maintain the tree for the rest of their lives. The city also provides family gardens for people without a garden area. In New South Wales, Australia, on the other hand, a couple Luke and Debbie have built a "360° rotating house" that can rotate to follow the sun and moves not on the basis of fossil fuels, but on electronic devices.

With forests and trees, Cittaslow can be greener. Today, we need to focus on the reduction of carbon emissions, thus burning CO_2, instead of earning. Spreading the movement to reduce burning instead of the earning movement guarantees a long future.

5.5 The Theory of Manbonism Before Capitalism

The Cittaslow movement, a happy community movement driven by slowness and freedom. The investment expert George Soros who is famous for his charity business wrote an autobiography entitled "The Philanthropy of George Soros: Building Open Societies." He mentioned in the 2012 Davos Forum, "The problem does not lie in capitalism, but in the malfunctioning democracy. It's not a crisis in capitalism, but in democracy" [111].

America, which was founded on the basis of democracy, is facing critical democracy issues, which hold back its economy. If the problems of capitalism can't be overcome, the rise of democracy problems is a natural result. The founder of the World Economic Forum (Davos Forum) Klaus Schwab confessed, "In the global intense competition, the social cohesion is missing in the pro-capitalism system, which is completely our responsibility." In Korea, 0.1 % of businesses add up 57 % of total business earnings in the country.

The fatal flaws in capitalism are the polarization of rich and poor, inequality, and financial greed just to name a few. According to the economist, Professor Jeffrey Sachs from Colombia University, global capitalism presents many direct threats to happiness and is weakening social trust and mental stability, with the prevalence of clinical depression on the rise. The multinational enterprises' thoughtless pursuit for profit is even raising the global temperature and destroying the natural environment.

Furthermore, global capitalism is weakening the social trust and mental stability. The American boxing champion Muhammad Ali's famous statement "Float like a butterfly, sting like a bee" can be closely related to this issue. Capitalism has encountered us with the beautiful figure of a butterfly, however, has ended up hurting us with a poisoned sting. This reminds of Vincent van Gogh's words, "I tell you, the more I think, the more I feel that there is nothing more truly artistic than to love people." In a society affected by speedholics, it is impossible to find human warmth. The current economy is terrifying, since there is no human element included.

The theme of the book *Plato: Republic* [112] by Plato, a teacher of Aristotle, tells what sort of activities enable one to live well (see Fig. 5.8). He explains that the soul distinguishes itself from the

Fig. 5.8 Plato (*left*) and Aristotle (*right*), who invented the theory of happiness in 4 BC (*Source*: en.wikipedia.org)

body and states that it is a shortcut to take care of the soul rather than the body for a beautiful life. As today we live in a society with a highly developed material civilization, we can forget that material is not everything and there is something more important that the above message reminds us of.

As it was continuously emphasized after Aristotle, a state's purpose is not merely to provide a living but to make a life that is good, and the purpose of a government is to enable its citizens to live a full and happy life. Therefore, it is crucial that the government should become an aggregation of experts on how to create happiness.

Since Aristotle, around 4 BC, the happiness index has been a high interest factor. Nevertheless, in Korea's presidential election of 2007, people have voted the former president with the 747 Plan,[8] which focuses on economic growth.

As soon as the government focused on economic criteria, such as the GDP, other issues, such as poor health condition, social trust

[8] The plan named after the Boeing 747 pledged to achieve 7 % economic growth, increase in GDP per capita income to $40,000, and to make the nation the 7th largest economy in the world.

decline, and environmental pollution, started to arise one after another, and these issues have affected the social happiness. The goal of politics has to become the establishment of happiness, freedom, and morals and ethics of capitalism. However, according to the survey of 1,000 Korean adults on who drives the society into misfortune, 67.5 % selected the politicians, followed by government officials (3.0 %), men from religious circles (2.6 %), and people in the legal profession (1.8 %), which implies that for many people politicians interfere in living a happy life [113].

Nonetheless, recently, there has been good news. The President of the European Council, Herman Van Rompuy, sent the book *The World Book of Happiness* to world leaders and urged them "to make people's happiness and well-being our political priority for 2012 (see Fig. 5.9)" [114]. The book is a "collection of short, factual essays on well-being and positive psychology by 100 of the world's leading happiness experts, picked from 50 different countries." The overall message of the book has been noted as, "Money can buy happiness, if we spend it on others." The reason for admiring the bee more than any other insect is not due to its diligence, but due to its

Fig. 5.9 The President of the European Council, Herman Van Rompuy, has suggested the happiness philosophy, "Money can buy happiness, if we spend it on others" through the book *The World Book of Happiness* (*Source*: en.wikipedia.org)

commitment to work for others. In the lyrics of John Lennon's famous song "Imagine," there is the following part:

Imagine
by John Lennon (1971)

Imagine no possessions
I wonder if you can
No need for greed or hunger
A brotherhood of man
Imagine all the people
Sharing all the world

Happiness is the good relationship with time, space, and humans. There are eight ways toward happiness: "count your blessings," "practice acts of kindness," "say our life's joys," "thank a mentor," "learn to forgive," "invest time and energy in friends and family," "take care of your body," and "develop strategies for coping with stress." The most important out of eight to live a more satisfactory life is to "invest time and energy in friends and family." The current time, the living space, and the meeting people that one encounters, all three have to be in harmony to become a happy person.

The majority of the felt happiness originates from good relationships, which are breaking if we live a busy life. It is vital to live in harmony with slownomics and Confucian capitalism, which emphasize relationship, tradition, and wisdom. These come from warmth and longing and slowness and creativity [115]. Even in the eight criteria to live a satisfying life, it states, "invest time and energy in family and friends," emphasizing the importance of relationships [116].

If we take a look at the book *Glück Kommt Selten Allein* (*Luck Rarely Comes Alone*) by the German physician and comedian Eckart von Hirschhausen, we need to consume "multigrain bread" in order to maintain the "density of happiness in the blood" [117]. Sugar is a critical food element for the brain. The human brain daily consumes 100 g of sugar. Although the brain only represents 3.5 % of the total body weight, it utilizes 20 % of the total energy. However, as multigrain bread steadily and durably maintains the density of happiness in

Fig. 5.10 The movie "The Lives of Others" (2006) asks us whether we live the life of others (*Source*: movie.naver.com)

the blood, the relationship between multigrain and happiness has been pointed out.

Happiness^{Manbonism} before economy^{capitalism}: Manboncapitalism. The Cittaslow movement aims to provide "freedom and happiness." There is a movie set in East Germany called "The Lives of Others" (2006) (see Fig. 5.10). The story talks about the following: monitoring and oppression of an artist, who enjoys freedom as his life. One of the scenes shows a National Security agent who has lived the life of a stranger after the collapse of the wall, whom "Die Sonate vom Guten Menschen (Sonata for A Good Man)" is dedicated to express thanks.[9]

[9] The movie "The Lives of Others" (2006)

Now is the time to think about how many people in this world are currently living the life of a stranger.

If we take a look at the story of an Indian old man, we can understand what it means to enjoy the simple and the small things in life and become happy.

An American from Chicago approached me and asked how much a pack of onions were.

The old man answered 10 cents.

Looking at the offer, the total amount that the old man was selling added up to around 20 packs.

Therefore, the American asked how much all 20 of them would be.

Although he was asking for a discount in return for purchasing in a mass quantity, the old man answered that he could not sell all of them at once.

The surprised American asked.

"Why can't you sell them all together?"

And the old man replied.

I came here to live my life.

I love this market. I love the bustling people around. I love the sun and palm trees moving in the wind.

Friends come along to have a talk.

I love their talk of their children and farming.

This is my life.

For this, I come and sit here all day to sell 20 packs of onions.

If I decide to sell them all at once, I might earn money

However, my day would be over.

Therefore, I would have lost the life I love.

This, I cannot do.

If the New Village Movement in the 1970s in Korea was built to revive the economy, the Cittaslow movement is established to revive the peoples' happiness and to build a warmhearted cultural economy in developed countries. The Cittaslow movement is not antagonistic to capitalism; instead, it provides a lifeline by which capitalism can be made more sustainable.

The global capitalism pressurizes us to make faster, work faster, consume faster, and live faster without having any preparation for the resulting aftereffect. The slow capitalism allows the economy to serve the people slowly but surely. Thus, in order to guarantee the existence of capitalism, Manbonism which can be referred to as slow capitalism needs to build the base and should be followed by capitalism. Figure 5.11 illustrates the Eastern character of "man," and it can be distinguished into "su" (water meaning liquid), "il" (day), "sa" (four), and "wu" (again). The four virtues, "slowness with curve," "free will," "ethical and moral soul," and "delight in creation," need to form the base of our everyday lives, just as the flowing water, in order to ensure a long-lasting and non-stopping flow of happiness.

Fig. 5.11 "Man" of Manbonism is just one of Eastern character and refers to our happiness that we are seeking. It has four main components including water (su), every single day (il), four virtues (sa), and repeatability (wu)

"The curvy slowness," the nature built by God, is curvy, whereas the city built by humans is straight. If slowness is curvy, then fast is straight. God's secret is hidden in slowness and curve (see Chap. 1).

God gave humans freedom, but didn't violate humans' free will to make their own choice. Freedom is crucial for happiness. Let's take a flower, for example, the eye is free to decide, whether to simply look at a flower or look at the beauty of the flower. Although everyone has the free will to decide between good and evil and between economic wealth and happiness, the consequences and suffering will be different. However, the people today sacrifice a large quantity of their time to work and improving the business without thinking about such issues. Recently, the Food and Drug Administration in Korea discovered a carcinogenic substance in Chinese chopstick production, which led to all sold items being recalled. In this case, a big business was involved with an overall turnover of USD 4.5 million. Although knowing about this issue, it continued to sell the chopsticks due to profit concerns. Even after the Food and Drug Administration exposed the business, it did not hand out any production-related documents. Although the continuous trading was a choice, there is no doubt that the company had chosen "evil." Many economists believe that economic growth and happiness are closely related. However, a sense of freedom is an absolute condition in happiness.

The word "economy" in Eastern characters is a shortened word with the meaning "administrating world and saving people." Is the current economy really one that controls the world and helps the people? Until now, the economy has solely focused on economic growth and development, sacrificing the peoples' quality of life. We have been neglecting the issue of common good such as the desired state of our society, the morals, ethical values, and justice. No matter how much the market economy develops, the common good is one of supremacy. If we want to extend the human society to be more lasting and sustainable, we need to define ethics and justice immediately. Just like a muscle, if the virtue is used repeatedly, it will grow stronger. The humans with virtues and ethics are animals with souls. There is a living world, which surpasses the material world such as the economy.

Whereas money is the base of capitalism, the theory of Manbon is contentment formed of slowness, freedom, soul, and delight. For humans, happiness is a fundamental requirement, which has a higher priority than money, and thus, happiness is defined as the fundamental concept of Manbonism. The philosophy of the Cittaslow refers to obtaining maturity instead of growth, quality of life instead of quantity of life, and the depth and dignity instead of speed. Korean culture is rooted in sinmyeong (merry mood), the Korean-style ecstasy, and its three elements are pathos (feeling, passion), logos (logic, reasoning), and ethos (ethics, soul). Thus, if we secure the roots of humanity, "sinmyeong" will be long lasting. In other words, if we secure the roots of humanity through pathos (heart), logos (mind), and ethos (heart, reasoning, inspiration), we can become happy and bring this excitement into the economy. There is no internal fulfillment in an economy without spirit, a CEO without a soul or a civil servant without a soul. The soul is "slow" in essence. In fact inspiration springs to mind of a man who has been preparing for a long time. God's secret is hidden in slowness and happiness. Let's lose the focus of the childish belief that materialism will satisfy the soul. We aim to build a Cittaslow, a better world, where everyone can live happily. We have the right to enjoy a happy life.

Bill Gates described the compliments for his efforts to help disadvantaged people through his business as creative capitalism. Gap, Hallmark, Dell, and Microsoft mark their products with a "RED logo," which refers to a part of their income to be given to aid charities. Furthermore, the Japanese company Sumitomo Chemical has had a big influence by sharing their own technology with a textile company in Tanzania to produce "mosquito nets" in bedrooms in order to help the prevention of malaria. By promoting their own company to be preferred to any other company, it provides a tipping point (small changes result in a bigger change) to ensure "creative delight."

The following example is a story out of Les Miserables. After it has been decided that Jean Valjean is innocent, the bishop tells him, who has just received freedom, "Jean Valjean my brother you no longer belong to evil. With this silver, I have bought your soul. I've ransomed you from fear and hatred, and now I give you back to God." Jean Valjean became just as the bishop said. The creative forgiveness of the bishop had completely changed his life through the use of creative delight.

The ABC principle of successful Cittaslow. In order for Cittaslow to be successful, it is essential to build the organic ecosystem based on three core elements called "ABC"; A stands for "administration (Cittaslow administration)," B for "business (cooperative enterprises)," and C for "communication (communication between administration and residents)."

A—Cittaslow administration. It is important to determine whether the Cittaslow policy is the highest priority or second/third highest priority in the Cittaslow area. Slow governance means becoming an open community through comprehensive management and administrative support for seven major (72 sub categories) categories including energy and environment policies, infrastructure policies, quality of urban life policies, agricultural, touristic and artisan policies, policies for hospitality, awareness and training, social cohesion, and partnerships. The administrative officer needs to transform the local community into a sustainable ecosystem. Cittaslow is just like a living organism, which is part of a circle of life (life, getting old, disease, and death).

B—Community business, cooperatives, aim to concentrate on the specialties of the local community with the outlook for the next 100 years that can create a lasting income and establish a revenue model that is self-growing and self-reliant and thus to create a community-based and resident-led business. We need to realize that relying solely on the subsidy of the government is doomed to fail. Cooperative market creation should lead to the creation of employment opportunities and create happiness. The scale of community businesses in England has reached USD 42.4 billion, whereas Japan aims to establish a market scale of USD 26.7 billion in the next 10 years.

C—Communication aims to prosper from the fruits of communication and consensus. The Cittaslow movement is an open community movement through communication between government and the Cittaslow Resident Council. It is also practicing an open community organic ecosystem through organic networking among Cittaslow member towns.

A concept is proposed for a successful Cittaslow. This "ABC" system seeks the triangular harmony of three elements, enabling a Cittaslow to be a "liquid Cittaslow" where governance, diversity, and communication continue to work like flowing water (see Fig. 5.12).

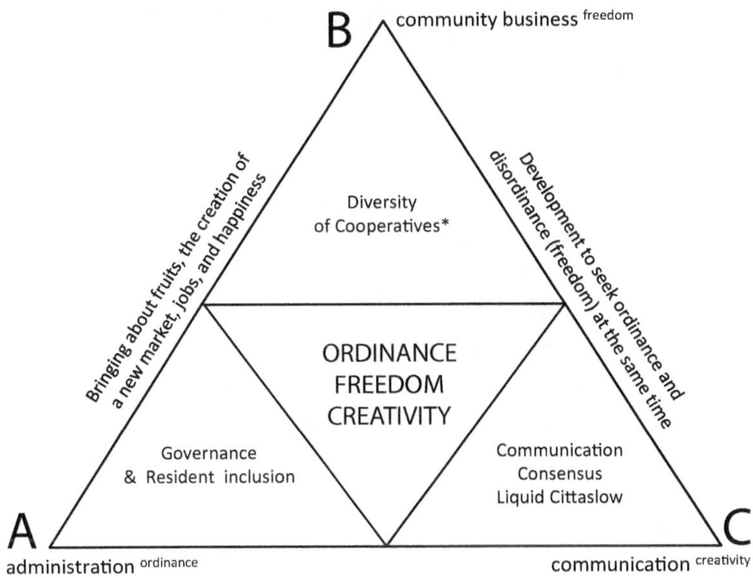

Fig. 5.12 Liquid Cittaslow System—triangular harmony. *Turning toward the more community-based localized economies rather than globalization creates more jobs and happiness. A, B, and C need to be liquid to work like flowing water

As "A" aims good ordinance and "B" pursues creative freedom, the system can be said to seek ordinance and disordinance at the same time to bring about the fruits, creation of a new market, jobs, and happiness. When free will in Manbonism is applied to a Cittaslow, more creative freedom is expected.

Communication is highly important. Although the New Village Movement[10] was the first industrialization movement in Korea, it was a movement controlled by the government. While the New Village Movement claimed to "pull people out of poverty," it resulted in the collapse of the agricultural community. On the contrary, the Cittaslow

[10] The New Community Movement, also known as the New Village Movement or Saemaul Movement, was a political initiative launched on April 22, 1970, to modernize the rural South Korean economy. The idea was based on the Korean traditional communalism called Hyang-yak and Doorae, which provided the rules for self-governing and cooperation in traditional Korean communities.

movement can strengthen the cohesion of the community. In this regard, Cittaslow and Slow Community are beautiful movements.

5.6 Points

Contemporary humans are suffering drastically from the endless obsession with money, making the current "economic person" into a production-consumption machine. God has given humans the free will to make choices. However, people have lost the balance and have chosen a materialistic lifestyle such as the economy, GDP, growth, real estate, energy, and competition as a priority.

Happiness happens; it's not gained. The origin of happiness is the soul, which is slow in essence. Cittaslow respects the original speed of nature. Happiness is the giving up of speed.

Religion and economy are the two elements that have made the biggest influence on humanity. Through the religious revolution in 1517, an important ideology has sprung up in Europe's modern history. Originally, wonderful economic ethics were created from the puritan capitalism. Korea, for example, has the Hwaeom idea based on "earning and sharing with others," which has evolved through the idea "Dasal" of Silla and has become a magnificent tradition.

Nature, as God's creation, is curvy, whereas the city made by humans is straight. The original route of a river is also naturally curvy. In the last 100 years, the average temperature in the world has risen by 0.74°, while in the Korean Peninsula the average temperature rise was noted as 1.5°, in six major cities up to 1.78°. If the temperature rises by 2°, it becomes difficult for people to live healthy. We need to build green arks in Cittaslow for nature through forests and trees. The sustainability of nature is the sustainability of humans.

The global capitalism is a direct threat to happiness, and there is no human warmth in capitalism. If facing a crisis, it is necessary to go back to the basics. We suggest a model of "Manbonism" developing from the principles of puritan capitalism and Buddhist economy. Whereas money is the base in capitalism, Manbonism is based on four virtues such as the curvy slowness, free will, a soul filled with virtues and ethics, and creative delight. That is, Manbonism is happiness. According to Plato, the good soul lives well and the unjust soul badly. This leads to the fact that a man is happy and profits from his justice; thus, the unjust man is miserable [118]. The ABC triangular harmony theory—Liquid Cittaslow System—was drawn from the theory of Manbonism applied to a Cittaslow. A Cittaslow can be successful when Manbonism is before capitalism and they cooperate with each other and are consilient.

References

1. sohn, d., & Jang, H. (2010). *Addicted to slow city* (pp. 16–17). Seoul, Korea: Chosun & Book.
2. Jang, S. (2005). *Slowness and emptiness* (p. 127). Seoul, Korea: Root and Leaf Publishing.
3. Leclercq, C. J. (2004). *Eloge de la pareses (Praise of laziness).* Korea: Bundo Books. In trans. of Jang, I., Waegwan.
4. Joseon Dynasty (1392–1897) is the last dynasty of Korean history.
5. UNESCO World Heritage Centre (2014). whc.unesco.org.
6. Bai Juyi (772–846), Chinese poet of the Tang Dynasty.
7. Shinichi, T. (2003). *Slow life* (p. 153). Korea: Design House. In trans. of Kim, H., Seoul.
8. Honoré, C. (2004). *In praise of slow*. London, UK: Orion Books.
9. Scheurmann, E. (1990). *The Papalagi: The speeches of Tuiavii from Tiavea* (p. 89). Korea: Mind Books. In trans. of Choi, S., Seoul.
10. Shinichi, T. (2003). *Slow life* (p. 164). Korea: Design House. In trans. of Kim, H., Seoul.
11. Honoré, C. (2004). *In praise of slowness: Challenging the cult of speed.* New York, NY: Harper One.
12. Schumacher, E. F. (2000). *Small is beautiful: A study of economics as if people mattered* (25th Anniversary Edition). London, UK: Hartley and Marks Publishers
13. Schumacher, E. F. (1998). *This I believe and other essays*. Cambridge, UK: Green Books.
14. Hesse, H. (2009). *Uber das Gluck von Hermann Hesse* (p. 13). Korea: Jong Publishing. In trans. of Oh, H., Seoul.
15. Bertrand, R. (2009). *Mortals and others* (p. 14). Oxon, UK: Routledge Classics.
16. Ban, C. (2005). *The power of laughing* (p. 17). Seoul, Korea: Poem & Poem.

d. sohn et al., *Go Slow and Curvy*,
DOI 10.1007/978-3-319-14854-0

17. Penn, M. & Zalesne, K. (2007). *Microtrends: The small forces behind tomorrow's big changes*. New York, NY: Twelve.
18. Godin, S. (2006). *Small is the new big: And 183 other riffs, rants, and remarkable business ideas*. New York, NY: Penguin Group.
19. Thoreau, H. D. (2008). *Walden*. Radford, VA: Wilder Publications.
20. Brûléd, T. (2010, November 13). *Brand Korea: Soft power, hard sell*. Financial Times
21. Hong, H. (2011). *Don't have two arrows in one hand at the same time. Happily together*. Seoul, Korea: Positive Thinking.
22. CIA. (2012). The World factbook. www.cia.go.
23. Bailey, E. T. (2010). *The sound of a wild snail eating*. Chapel Hill, NC: Algonquin Books.
24. Honore, C. (2010, August 29). *Keynote speech in the 2010 World Leisure Congress*. Korea.
25. William, P. (2010). *Hamlet's blackberry*. New York, NY: Harper Collins.
26. Euromonitor International (2012). *Special report: Rise in single-person households globally impacts consumer spending patterns*. Euromonitor International.
27. Eurostat. (2012). *Statistics explained archive*, Vol. 2 – Social statistics.
28. Statistics Korea. (2011). *Population and housing census 2010*. Korea: Statistics Korea.
29. UN DESA. (2011). *World urbanization prospects*, the 2011 revision. The United Nations Department of Economic and Social Affairs.
30. The Korean Ministry of Land, Infrastructure and Transport. (2008). *Report: A Comprehensive Plan to Climate Changes*. The Korean Ministry of Land, Infrastructure and Transport.
31. World Bank (2011). www.worldbank.org.
32. sohn, d. (1998). *Funship*. Seoul, Korea: Hyungseul Publishing.
33. BP Global. (2010). *Statistical review of world energy 2010*. BP Global.
34. Root, T. L., Price, J. T., Hall, K. R., Schneider, S. H., Rosenzweig, C., & Pounds, J. A. (2003). Fingerprints of global warming on wild animals and plants. *Nature, 421*, 57–60.
35. Chungbuk Forest Environment Research Institute, Korea (2011).
36. Lee, S., Ellwood, E. R., Park, S., & Primack, R. B. (2011). Late-arriving barn swallows linked to population declines. *Biological Conservation, 144*, 2182–2187.
37. Shinichi, T. (2003). *Slow life* (p. 125). Design House: Korea. In trans. of Kim, H., Seoul.
38. sohn, d. (2000). *The charm of Korea and travel* (p. 252). Seoul, Korea: Baeksan Publishing.
39. Hamel, G. (2009). Presentation: Ultimate advantage: Reinventing management for a new age. *The World Knowledge Forum 2010*.
40. Park, D. (2011). The secret of circus king. *Chosun Daily*, Korea, 2011. 5. 29.

41. Lee, J. (2012). Modern business success key: 'agility'. *Chief Executive*, Issue June.
42. Pink, D. (2004). The MFA is the MBA, Breakthrough Ideas for 2004. *Harvard Business Review, 21.*
43. sohn, d., & Jang, H. (2010). *Addicted to slow city* (p. 82). Seoul, Korea: Chosun & Book.
44. Lee, S. (2011, July 5). Naked Russian diver Natalia Avseenko and dolphins. *Joongang Daily*, Korea
45. Shinichi, T. (2003). *Slow life* (p. 34). Seoul, Korea: Design House. In trans. of Kim, H.
46. The word of 'Sunim' refers to 'Korean Buddhist monk'.
47. Hyeongak (2010,). Monk Hyeongak's sermon. *The 2010 Cittaslow International Assembly Corea*. 2010. 6. 28.
48. Honore, C. (2004). *In praise of slowness: Challenging the cult of speed.* New York: Harper One.
49. Shinichi, T. (2003). *Slow life* (p. 125). Seoul, Korea: Design House. In trans. of Kim, H.
50. The U.S. Department of Health and Human Services (2011). A profile of older Americans.
51. The Japan Ministry of Internal Affairs and Communication (2014). http://www.stat.go.jp/.
52. Shinichi, T. (2003). *Slow life* (p. 187). Seoul, Korea: Design House. In trans. of Kim, H.
53. sohn, d. (2005). *Culture is a reborn business by entertainment industry.* Seoul, Korea: Gimmyoung Publishers.
54. Leclercq, C. J. (2004). *Eloge de la pareses (Praise of laziness).* Waegwan, Korea: Bundo Book. In trans. of Jang, I.
55. Gates, B. (2008, August 11). How to fix capitalism. *Time.*
56. Lee, J., Song, S., & Shin, H. (2011, March 29). The good infrastructure is a national power. *Maeil Daily*, Korea.
57. sohn, d. (2008, 27 July). Make the people and employees dance! *Chosun Weekly Biz*, Korea.
58. Schuster, T. (2011, September 24). The energy of employees is the strength of the enterprise. *Chosun Weekly Biz*, Korea.
59. Lee, J. (2010). *Do you have soul, creativity, and communication?* Seoul, Korea: Sam & Parkers.
60. DeMarco, T. (2010). *Slack: Getting past burnout, busywork and the myth of total efficiency.* Seoul, Korea: Insight Books. In trans. of Ryu, H., Lee, B., & Hwang, J.
61. sohn, d. (2008, July 27). Make the people and employees dance! *Chosun Weekly Biz*, Korea.
62. Lee, S. (2011). *Sereno leaders.* Seoul, Korea: Salt Tree Publishing.
63. Jang, D. (2011, September 22). An unstable period needs soft leadership. *JoongAng Daily*, Korea.

64. Choi, B. (2011, February 28). All tiny stitching by a master is art. *Chosun Daily*, Korea.
65. Korea Broadcasting Station. (2011, November 2011). One hundred-year-old businesses.
66. sohn, d. (2011, August 20). The real goal of slow movement and education is human happiness. *Daily University News Network*, Korea.
67. The Economist (2012, March 3). Small is not beautiful.
68. The Economist (2012, March 3). Decline and small.
69. Hangyere Daily (2013, October 20). In an interview with Christian Kellermann.
70. Borscheid, P. (2008). *The tempo virus: A cultural history of the acceleration.* In trans. of H. Doo. Seoul, Korea: Deolneok Publishing.
71. Urban Agriculture Helps Combat Hunger in India's Slums. (January 17, 2013). www.worldwatch.org
72. Im, S. (2009). *A simple & plain life: The lesson from the Amish village.* Seoul, Korea: Leesu.
73. Maeil Daily. (2010). *Agriculture needs to be an export industry like semiconductors and cars.* Korea, March 24, 2010.
74. The United Nations, Department of Economic and Social Affairs. (2012). *World urbanization prospects*, the 2011 Revision.
75. The Korea Ministry of Land, Transport and Maritime Affairs. (2008). *The master plan on climate changes.*
76. The World Bank. (2011). *Online data.* www.wolrdbank.org
77. The National Institute of Environment Research. (2012). *A press release.* The Korea Ministry of Environment.
78. The UN Commodity Trade Statistics Database. (2011). http://comtrade.un.org
79. Leibinger, R. (2011). *A speech at the Cittaslow coordinating committee in Waldkirch, Germany.* September 3, 2011.
80. Jeong, J. (2011). The desert feeds Israel. *JoongAng Daily, Korea.* September 9, 2011.
81. Honore, C. (2004). *In praise of slowness.* Canada: Knopf Canada.
82. Cittaslow Mold. (2011). *Mold food and drink festival edition.* September, 2011.
83. Ahn, H. (2011). A proper understanding of Korean beans. A special lecture on the Cittaslow club 'Jihe,' *The National Cittaslow Corea Network*, August 25, 2011.
84. Go, E. (2011). *Beans and traditional Korean fermented bean sauces.* Cittaslow Namyangju, Slow Food Korea.
85. Schumacher, E. F. (1997). *This I believe and other essays.* Totnes, UK: Resurgence Book.
86. Schumacher, E. F. (1973). *Small is beautiful: A study of economics as if people mattered.* London: Blond & Briggs.

87. sohn, d., & Jang, H. (2010). *Addicted to slow city*. Seoul, Korea: Chosun & Book.
88. Joongang Daily. (2008). *The boom of new human cities in the world.* Korea, January 22, 2008.
89. Alysham Town Council. http://aylshamtc.norfolkparishes.gov.uk/
90. Hosouchi, N. (2007). *Community business*. In trans. of H. Park, & S. Lee. Seoul, Korea: Arche Publishing House.
91. Lee, W. (2011). Let' make Seoul a town! *Hankyoreh Daily*, Korea, October 26, 2011.
92. Jang, D. (2012). Salt will keep you healthy. *Jellado.com*, August 14, 2012.
93. Lee, H. (2010). Soft cities in the world: Orvieto, Italy. *Hankook Daily*, Korea, November 13, 2010.
94. Choi, B. (2011). Spitalfields Market, the best in England. *Chosun Daily*, Korea, July 24, 2010.
95. Kim, S. (2011). Interview with Amy Cortese, author of 'Locavesting: The revolution in local investing and how to profit from it.' *Chosun Weekly Biz*, Korea, December 3, 2011.
96. Diener, E. (1984). Subjective well-being. *Psychological Bulletin, 95*(3), 542–575.
97. Weiner, E. (2010). *The geography of bliss*. In trans. of S. Kim. Seoul, Korea: Woongjin ThinkBig.
98. Kahneman, D., & Deaton, A. (2010). High income improves evaluation of life but not emotional well-being. *Proceedings of the National Academy of Sciences of the United States of America, 107*(38), 16489–16493.
99. The Institute of Social Development and Policy Research. (2011). *The future Korea report 2011 forum.*
100. Beopjeong, M., & Ryu, S.-H. (2006). *Be happy, all living creatures!* Seoul, Korea: Harmonious Life.
101. Gil, H. (2006). *Spirituality. A speech at Saegil Church, Korea*, August 19, 2006.
102. Cha, D. (2009). Sermon on the Mount – the eight beatitudes. *JoongAng Daily*, Korea, November 5, 2009.
103. sohn, d. (2000). *The charm of Korea and travel* (p. 43). Seoul, Korea: Baeksan.
104. Weber, M. (1985). *The protestant ethic and the spirit of capitalism.* Die Protestantische Ethik und der Geist des Kapitalismus, In trans. of S. Gwon, & M. Kim. Seoul, Korea: Iljogak Publishing.
105. Song, B. (1987). *Emotional economics* (pp. 138–155). Seoul, Korea: Parkyoungsa.
106. sohn, d. (1998). *Funship* (p. 172). Seoul, Korea: Hyungseul.
107. Chrysostom, J., Edward, B., & Charles, M. (1842). *Homilies concerning the statues to the people of Antioch Homilies of St. John Chrysostom.* Oxford: J.H. Parker.
108. Thoreau, H. D. (2004). *Walden.* Princeton, NJ: Princeton University Press.

109. Germany: Nuclear power plants to close by 2022. (2011 May, 30). www.bbc.co.uk
110. Park, J. (2011). Interview: Thirawit Leetavorn, senior executive vice president of 'Double A.' *JoongAng Daily*, Korea, November 12, 2011.
111. Choi, W. (2012). It is up to the government to make the people happy. *Chosun Weekly Biz*, Korea, February 4, 2012.
112. Plato, Grube, G. M. A., & Reeve, C. D. C. (1992). *Plato: Republic* (2nd ed.) Cambridge, MA: Hackett.
113. Lee, J. (2011). Poll cites politicians as root of social unhappiness. *Yonhap News Agency*, Korea, December 1, 2011.
114. Lee, S. (2011). The EU president's gift, 'World Book of Happiness.' *Chosun Daily*, Korea, December 24, 2011.
115. sohn, d., (2009). Slownomics for sustainable growth. *Chosun Weekly Biz*, Korea, October 24, 2009.
116. Wallis, C. (2005). The new science of happiness. *Time*. January 17, 2005.
117. Hirschhausen, E. (2010). *Happiness is not a matter of luck*. Glück Kommt Selten Allein. In trans. of G. Park (pp. 227–228). Seoul, Korea: Gingko Trees Publishing.
118. Plato, Grube, G. M. A., & Reeve, C. D. C. (1992). *Plato: Republic*. Indianapolis, IN: Hackett.

www.ingramcontent.com/pod-product-compliance
Lightning Source LLC
Chambersburg PA
CBHW071339230926
51846CB00002B/386
* 9 7 8 3 3 1 9 3 7 2 1 1 2 *